PERCEPTION, REALISM, AND THE PROBLEM OF REFERENCE

One of the perennial themes in philosophy is the problem of our access to the world around us: do our perceptual systems bring us into contact with the world as it is, or does perception depend upon our individual conceptual frameworks? This volume of new essays examines reference as it relates to perception, action, and realism, and the questions which arise if there is no neutral perspective or independent way to know the world. The essays discuss the nature of referring, concentrating on the way perceptual reference links us with the observable world, and go on to examine the implications of theories of perceptual reference for realism, and the way in which scientific theories refer and thus connect us with the world. They will be of interest to a wide range of readers in philosophy of science, epistemology, philosophy of psychology, cognitive science, and action theory.

ATHANASSIOS RAFTOPOULOS is Professor of Epistemology and Cognitive Science in the Department of Psychology at the University of Cyprus. He is the author of *Cognition and Perception: How Do Psychology and the Neural Sciences Inform Philosophy* (2009), editor of *Cognitive Penetrability of Perception: Attention, Action, Planning, and Bottom–Up Constraints* (2005), and co-editor, with Andreas Demetriou, of *Cognitive Developmental Change: Theories, Models and Measurement* (Cambridge, 2011).

PETER MACHAMER is Professor of History and Philosophy of Science at the University of Pittsburgh, Associate Director of Pitt's Center for Philosophy of Science, and a member of the Center for the Neural Basis of Cognition. He is co-author, with J. E. McGuire, of *Descartes' Changing Mind* (2009). He is co-editor, with Gereon Wolters, of *Interpretation* (2010); and, with Michael Silberstein, of *Blackwell's Guide to Philosophy of Science* (2002).

PERCEPTION, REALISM, AND THE PROBLEM OF REFERENCE

EDITED BY

ATHANASSIOS RAFTOPOULOS

AND

PETER MACHAMER

CAMBRIDGE UNIVERSITY PRESS

CAMBRIDGE
UNIVERSITY PRESS

University Printing House, Cambridge CB2 8BS, United Kingdom

Published in the United States of America by Cambridge University Press, New York

Cambridge University Press is part of the University of Cambridge.

It furthers the University's mission by disseminating knowledge in the pursuit of
education, learning and research at the highest international levels of excellence.

www.cambridge.org
Information on this title: www.cambridge.org/9781107414648

© Cambridge University Press 2012

First published 2012
First paperback edition 2014

A catalogue record for this publication is available from the British Library

Library of Congress Cataloguing in Publication data
Perception, realism and the problem of reference / [edited by] Athanassios Raftopoulos,
Peter Machamer.
p.   cm.
Includes bibliographical references and index.
ISBN 978-0-521-19877-6 (hardback)
1. Reference (Philosophy)   2. Perception (Philosophy)   3. Realism.
I. Raftopoulos, Athanassios.   II. Machamer, Peter K.
B105.R25P46   2012
121–dc23   2011051918

ISBN 978-0-521-19877-6 Hardback
ISBN 978-1-107-41464-8 Paperback

Contents

v

Contents

Figures

Contributors

WILLIAM BECHTEL, Professor of Philosophy, University of California, San Diego

AMANDA BROVOLD, PhD student, University of California, San Diego

DEREK H. BROWN, Associate Professor of Philosophy, Brandon University

JOHN CAMPBELL, Professor of Philosophy, University of California, Berkeley

RICK GRUSH, Professor of Philosophy, University of California, San Diego

DON HOWARD, Professor of Philosophy, University of Notre Dame

E. J. LOWE, Professor of Philosophy, University of Durham

PETER MACHAMER, Professor of Philosophy, University of Pittsburgh

MOHAN MATTHEN, Professor of Philosophy, University of Toronto

LISA OSBECK, Associate Professor of Philosophy, University of West Georgia

DEMETRIS PORTIDES, Assistant Professor of Philosophy, University of Cyprus

STATHIS PSILLOS, Professor of Philosophy, University of Athens

ATHANASSIOS RAFTOPOULOS, Professor of Epistemology and Cognitive Science, University of Cyprus

GERALD VISION, Professor of Philosophy, Temple University

CHAPTER I

Reference, perception, and realism

Athanassios Raftopoulos and Peter Machamer

One of the perennial themes in philosophy has been the problem of our access (if any) to the world around us. This is, widely construed, the problem of realism. The question can be made more specific. Do our perceptual systems bring us into contact with the world as it is or do some of the ways they present the world depend on the systems themselves, e.g. as due to one's conceptual frameworks or to the make up of the perceptual systems? If they do not show us the world as it really is, and if there is no neutral perspective or independent way to know the world, does it make sense to talk about a world that exists independently of organisms that perceive or experience it? A somewhat related question is whether our scientific theories reveal the world as it really is or whether the theoretical assumptions and concepts constitutive of every theory somehow describe the world only in ways presupposed by the theory and its background assumptions.

The answers to these questions hinge in the last analysis on whether our perceptual acts, such as fixing the eyes and other deictic operations, succeed in picking out real objects or features in the world. Since perception is the basis for our evidence for scientific theories, the related question becomes whether the terms, especially the theoretical terms, of our best scientific theories ought to be taken to refer to entities, events, and processes in the world. If they do, then our perceptual system and our best scientific theories would correctly represent the world. Of course, scientific theories have a history, and, so even at best, they do not always get things right.

One might think that to address these questions one should be able to stand outside any perceptual system and any theory, and see from that neutral, objective standpoint the way perceivers see the world and the ways theories depict or represent the world. Then one could judge whether perception and theories deliver the world faithfully or accurately.

This position outside perceptual systems and theories would be a metaphysical Archimedean point (Kitcher 2001) from which one could compare our representations of the world, whether through our perceptual systems or through our theories, and the (allegedly) mind-independent world we represent. This objective point of view does not exist, and if Putnam (1981, 1982) and Rorty (1980) are right, no one could ever find such a standpoint and answer definitely the above-mentioned questions by showing that our perception and scientific theories depict the world accurately. The reason is that the only way one could determine the reference of perceptual demonstratives and of scientific theoretical terms is through the perceptual system and our best available scientific theories. It seems metaphysically impossible that we could ever directly, and independently of the perceptual system and of our theories, compare our symbols with the world they represent; inevitably information about the world is delivered to us through our perceptions and through our theories.

Considerations like these have given rise to two powerful attacks on realism from realism's nemesis – to wit, constructivism. Constructivists claim that any knowledge of material objects is constructed out of representations, and that the objects of these representations, as mind-independent entities, are epistemically inaccessible. Constructivism denies the realist's claims that scientific theories tell us about mind-independent objects.

Epistemological constructivism undermines realism by arguing that our experience of the world is mediated by our concepts, and that there is no direct way to examine which aspects of objects belong to them independently of our conceptualizations. There is no metaphysical Archimedean point from which one could compare our representations of objects and the mind-independent objects we represent. Perception is cognitively penetrable and theory laden. More specifically, the thrust of constructivism's argument is that the theory-ladenness of our perception implies that our experience is mediated by our concepts, and thus:

(a) People with two different conceptual backgrounds experience the world differently and may refer to different entities or processes even when they view the same scene.

(b) They could agree on what they see only if they had the same conceptual framework.

(c) There could be no theory-neutral basis on which debates about theory testing, confirmation, and choice could eventually be resolved. From this ensues the famous incommensurability thesis that bars communication across paradigms. It also follows that if there is no neutral basis on which to decide whose references are successful and whose are not, for the same (allegedly) visual scene, it may not make sense to talk about there being one visual scene that two people perceive.

Epistemological *constructivism* can be traced back to N. R. Hanson's *Patterns of Discovery* (1958) and Quine's famous 'gavagai problem', which Quine (1960) developed to argue from the indeterminacy of translation to the indeterminacy of meaning and reference. In Quine's (1970) later work, the argument is extended from linguistic utterances to mental states.[1] It can also be readily detected in the undermining of the theory-neutrality of perception, which has rendered the distinction between *seeing* and *seeing as* obsolete (Churchland 1988; Hanson 1958; Kuhn 1962), clearing the way for theories of science and meaning as historically relative. Since the existence of a theory-neutral basis for a rational choice among alternative theories was rejected, it was held that scientific theories are incommensurable.

Semantic constructivism attacks realism on the ground that there is no direct way to set up the relation between the terms of representations and the entities to which they purportedly refer. That relation can only be indirect, mediated through causal relations between these entities and our behavior. The relation can only be interest dependent. Since these relations ground terms or representations by fixing their referents, reference becomes theory dependent (Brandom 1996).

Note that constructivism's theses entail that one could not in principle ever know whether perceptual demonstratives and scientific terms refer to real entities and features in the world. They are mute as to whether there exists a mind-independent world. Some constructivists could be indirect epistemological realists, that is, they could hold that the immediate objects of perception are always (or at least typically) mental experiences. On this view, any perceptual access to the external or mind-independent world is robustly indirect.

It is in this way that the notions of reference, perception, and realism become interwoven. For realism to fight back, realists must undermine both epistemological and semantic constructivism. Against epistemological

---

[1] 'To expect a distinctive physical mechanism behind every genuinely distinct mental state is one thing; to expect a distinctive mechanism for every purported distinction that can be phrased in traditional mentalistic language is another. The question whether the foreigner really believes *A* or believes rather *B*, is a question whose very significance I would put into doubt. This is what I am getting at in arguing the indeterminacy of translation' (Quine 1970, 180–181).

constructivism, realists must show that perception, or some stage of it, is not theory laden. Since theories are broadly construed as conceptual frameworks, the realists must show that perception (or some part of it) is not conceptually mediated, that is, that it is conceptually encapsulated. More specifically, they must show that some of the contents of perception are not affected by the concepts that the perceiver may possess. Since, in the brain, contents are abstract entities carried by neural/mental vehicles, the above requirement is transformed into that of showing that the perceptual states are not modulated by conceptual/cognitive states, which, in cognitive-science parlance, means that they are not cognitively penetrated. Note that if realists succeed in this task, they will show that some perceptual pick-up of the world does individuate and track objects and features in the environment. This would be tantamount to saying that perception can secure an interest-free reference to some aspects of the world, most likely to those aspects that are important to our species and which our perceptual systems have evolved to pick up. Thus, in arguing for the theory-neutrality of perception, realists also argue against a part of semantic constructivism.

To bring their argument home, realists must examine closely the relevant scientific evidence, because it is empirical argumentation, and not philosophical speculation, that could determine whether perception or some part of it is conceptually encapsulated or cognitively impenetrable. It is also scientific research that will show what information can be represented during that stage of perceptual processing. This is important because realists have to surmount Sellars' (1956) critique of the 'myth of the given', and Churchland's (1988) view that even if there is some rigidity and theoretical neutrality at an early stage of the perceptual process, this 'pure given', or sensation, is useless in that it cannot be used for any 'discursive judgment', since sensations do not have truth-values, and are not semantically contentful states. Only 'observation judgments' can do that, they claim, because they have content. Their content is a function of a conceptual framework and, hence, such judgments are theory laden. Thus, realists must show that the conceptually encapsulated content of perception is rich enough to be epistemologically interesting, that is, to allow them to build their case against constructivism and relativism by showing, for example, how discursive judgments, which constitute unarguably the empirical data used in testing and evaluating scientific theories, are constrained by the content of the theory-neutral stage of the perceptual process. This, in turn, would require both an account of how one comes to know the referents of her perceptual demonstratives and an account of the process of conceptualization that eventually takes place and enables perceptual judgments (for it is one thing to be able to individuate and track objects and features in the environment and another thing to be able to know these objects and features). Finally, science is important to showing whether the cognitively impenetrable or conceptually encapsulated content of perception is content at the personal level, that is, content of which the perceiver can be aware, or at the subpersonal level.

However, it is not enough for the realist to show that perception or some part of it is cognitively impenetrable to top–down conceptual influences. Perception may be conceptually modulated and, thus, theory laden, even if it is not penetrated by top–down cognitive/conceptual effects, by having concepts built into the perceptual system itself. To solve the problem of the underdetermination of the distal object and percept by the retinal image, our perceptual system employs a set of hard-wired principles reflective of the geometry and the physics of our environment (see Spelke's 1990 'object principles', Burge's 2010 'formation principles', and Raftopoulos' 2008, 2009, 'operational constraints'). Since the contents of these principles consist of concepts, arguably perception inherently contains concepts and, thus, even though it is not affected by the concepts in the higher cognitive systems, it is conceptually structured. In addition, at least some of the processing principles reflect some sort of 'theory' about the world that our perceptual systems have constructed in their evolutionary development to cope successfully with their environment.

A realist might also wish to explore another position that could complement the aforementioned strategy and which would draw support from a recent revival of the old Gibsonian theory of direct or ecological perception, according to which our perceptual systems retrieve all the information they need so that we could interact with the environment in realistic situations (hence the term 'ecological') in a direct, cognitively, and conceptually free way. The recent revival is often described under the terms 'situated' and 'embodied' perception, that is, the view that perception is not, primarily, a passive contemplation of the world but an active engagement with it by persons, with bodies, who act always within a specific situation and with specific needs and purposes. This brings the role of action into the picture, and paves the way to new strategies to explore the ways perceptual demonstratives have their reference fixed. At a first glance one might think that this is a thorny road for the realist to take, since appealing to actions on the environment that aim to satisfy the needs of an organism seems to render reference-fixing all the more dependent on the interests of the organism. However, this would overlook the fact that our on-line interactions with the environment are effected by our dorsal system, which functions entirely independently of any cognitive/conceptual interference. It also neglects the fact that many of our interests are natural and objective relative to certain environments. If the realist could show how perceptual reference could also be fixed through the interplay and coordination of perception and action that takes place along the dorsal system she would have gained considerable ground. Furthermore, in view of the fact that the dorsal system mediates our on-line immediate interactions with the environment, and that the processing along the dorsal path is not modulated in any way from top–down cognitive information, the realist has the opportunity to argue from the success of an organism's action to the adequacy of the representations formed in the dorsal system. The dorsal system processes information for guiding actions and locating objects in space.

It goes without saying that the realist would have to accept that different organisms with different needs – and, thus, with perceptual systems that have evolved differently – would cut up the world differently (cf. Letvin et al. 1959). The view that species-different organisms cut the world at different junctures – that is, the view that in perception only certain features are selected and that this selectivity is an inherent part of perception as it has been shaped by evolution and learning – does not imply that perception depends on conceptual schemes and, thus, that it is theory laden. It just suggests, in Vision's (1998, 411) words, 'that we [different species] are certain kinds of information processing engines and not others'.

Should the realist succeed in these initial moves, she would have also answered semantic constructivism, for she would have shown how conceptually unmediated reference to observable entities and their features is possible. However, this does not by itself answer the other semantic-constructivist thesis, namely the claim that the relation between the theoretical terms of our scientific theories and the entities to which they purportedly refer must depend on our theorizing and ground the theoretical terms in the entities to which they refer by fixing their referents. Thus, reference becomes theory dependent.

The situation gets more complicated for the realist if she takes into account the fact that our theories consist essentially of models of reality and, as such, are abstracted from, and partial descriptions of, the real systems that the theories purport to explain. As such, they are idealizations of the real world. Being idealizations, they naturally give rise to the question, to what sort of entities or processes do the theoretical terms of the sciences refer? To what entities do the phrases 'harmonic oscillator', 'ideal gass', or various constructs in the models of nuclear physics, refer?

The strategy that the realist had to follow to rebut epistemological constructivism is not available to her in this case. For if it is conceivably possible to show that perceptual demonstratives can refer directly, meaning without conceptual mediation, to entities in the world, it is unquestionable for the realist that the burden of reference of theoretical terms rests with the theory and its concepts, in the sense that what they refer to is largely determined by the theory in which they feature. Furthermore, some theoretical terms are transtheoretical in the sense that they can refer to the same entity even though they may occur in different theories. Similarly, different terms in the same or different theories may refer to the same entity. Even though these two conditions have an inherent, minimal realistic flavor, it is no easy task for the realist to construct a realist theory of reference, given the two predominant 'models' of reference-fixing, to wit, the descriptivist and the causal theories of reference. The problem is that both views of reference-fixing face well-known problems that largely stem from their respective demands that pure descriptions or bare causal chains should be sufficient to ground reference.

The task of the realist would be to examine the assumptions underlying these two theories of reference, and try to figure out a way to overcome their difficulties by revising or undermining the underlying assumptions. Having revised some of the assumptions, the realist might devise a new theory of reference that keeps the strengths of the two theories and leaves out their weaknesses. This usually presupposes that the new theory would be a combination of the two models, say a causal descriptivism. At this juncture the strategies open to semantic and epistemological realism may intersect. Raftopoulos and Muller (2006) and Raftopoulos (2009) have proposed such a causal 'descriptivist' theory for reference-fixing of the perceptual demonstratives that purports to render the fixing of the referents of these demonstratives direct, that is, conceptually unmediated and interest free, while evading the problems inherent to pure causal accounts of reference – most prominently the problem of which element of the relevant causal chain is the referent and the problem of explaining referential failure or misrepresentation. It claims to succeed in this by appealing to the non-conceptual content of perceptual demonstratives, which consists predominantly in spatio-temporal information, that acts as a 'description' that picks out the referent of the demonstrative. But it does not constitute a 'semantic fact' that allows it to function as a way to determine the same referent across perceptual contexts, as descriptions are supposed to do in the Fregean tradition. In that tradition, whatever satisfies the description associated with a singular or predicate term is among the referents of the singular or predicate term, since the salient description is strictly context dependent and cannot function outside the specific context that has created it. That is, the content of the mental act of perceptual demonstration is idiosyncratic to the relationship of the viewer with the visual scene, which means that different viewers may use different information to parse a scene or that the same viewer may use different information to individuate the same objects, depending on the viewer's perspective on the scene. This entails that the *de re* relationship of the perceiver with a visual scene (a relationship that allows her to retrieve information about the scene from the scene itself and not from a description of it) is highly contextual. This, in turn, means that a *de re* perceptual mode of presentation determines reference given or within a certain context.

To put it differently, one should be careful to distinguish between a description used to individuate and track an object in a visual scene and a description used semantically to fix the referent of the relevant mental perceptual demonstrative, thus allowing someone with the same information to individuate the same object just by acquiring this information and without perceiving the scene. In perception, the former is certainly the case, not the latter. To be able to individuate the same objects on viewing the scene, another viewer is not required to have or understand anything about the information used by the first viewer to individuate the same objects in that visual scene; other information may be used, depending on the idiosyncratic relationship between the viewer and the referent of the perceptual demonstrative, since the list of properties that allow object individuation in a visual scene is heterogeneous and may differ from case to case. It is in this sense that the information used to individuate and track objects does not constitute a 'semantic' description of the referendum.

Although Campbell's account of demonstrative reference differs from Raftopoulos' (2009) and Raftopoulos and Muller's (2006) account in some important ways, Campbell argues for a similar solution to the problem of fixing the referents of perceptual demonstratives. Campbell insists that spatial and motion information about an object together constitute its mode of presentation in a perceptual demonstrative and that this mode or sense fixes the reference of the demonstrative by drawing attention to the object. He also insists that one should not associate the sense of the demonstrative with a description of the object's features and location. The role of location consists in providing the binding parameter for singling out objects, and not in providing some sort of descriptive identification of the object. Location organizes the information-processing procedures that process information about that object. It is in this sense that 'the description completes the character of the associated occurrences of "dthat" but makes no contribution to content. It determines and directs attention to what is being said. ... The semantic role of the description is pre-propositional; it induces no complex, descriptive elements to content' (Campbell 2002, 107).

Yet another way to fix reference without relying on conceptual individuation would be to do so by acting or behaving in specific ways with respect to an object in the immediate environment. In this case the actions of an agent or the behavior of an organism would fix the referent by an appropriate activity, e.g. grasping the object, touching it, or, as in the case of children, by shared gazes towards an object or event that is already individuated by standing out against a background. Indeed, pointing at an object, in the right context, may suffice to establish a referent.

The chapters in this volume address some of the problems discussed in this brief introduction. Some of them focus on the problem of the reference in perception and on the problem of the role of action in reference-fixing. The rest focus on the reference of the theoretical terms of scientific theories.

REFERENCES

Brandom, R. B. (1996) *Making It Explicit: Reasoning, Representing, and Discursive Commitment.* Cambridge, MA: Harvard University Press.

Burge, T. (2010) *Origins of Objectivity.* Oxford: Clarendon Press.

Campbell, J. (2002) *Reference and Consciousness.* Oxford: Clarendon Press.

Churchland, P. M. (1988) Perceptual plasticity and theoretical neutrality: A reply to Jerry Fodor. *Philosophy of Science* 55: 167–187.

Hanson, N. R. (1958) *Patterns of Discovery.* Cambridge University Press.

Kitcher, P. (2001) Real realism: The Galilean strategy. *Philosophical Review* 110, no. 2: 151–199.

Kuhn, T. S. (1962) *The Structure of Scientific Revolutions.* University of Chicago Press.

Levin, J. Y., Matturana, H. R., McCulloch, W. S., and Pitts, W. H. (1959) What the frog's eye tells the frog's brain. *Proceedings of the IRE* 47, no. 11. Reprinted in W. C. Corning and M. Balaban (eds.) *The Mind: Biological Approaches to Its Functions.* New York: Interscience Publishers, 1968, pp. 233–258.

Putnam, H. (1981) *Reason, Truth and History.* Cambridge University Press.

(1982) Why there isn't a ready-made world. In *Realism and Reason*, vol. III of *Philosophical Papers.* Cambridge University Press.

Quine, W. V. O. (1960) *Word and Object.* Cambridge, MA: MIT Press.

(1970) On the reasons for indeterminacy of translation. *Journal of Philosophy* 67: 178–183.

Raftopoulos, A. (2008) Perceptual systems and realism. *Synthese* 164: 61–99.

(2009) *Perception and Cognition: How do Psychology and the Neural Sciences inform Philosophy*. Cambridge, MA: MIT Press.

Raftopoulos, A. and Muller, V. (2006) Nonconceptual demonstrative reference. *Philosophy and Phenomenological Research* 72, no. 2: 251–285.

Rorty, R. (1980) *Philosophy and the Mirror of Nature*. Princeton University Press.

Sellars, W. (1956) Empiricism and the philosophy of mind. In H. Feigl and M. Scriven (eds.) *Foundations of Science and the Concepts of Psychology and Psychoanalysis*. Minnesota Studies in the Philosophy of Science, vol. 1. Minneapolis: University of Minnesota Press.

Spelke, E. S. (1990) Principles of object perception. *Cognitive Science* 14: 29–56.

Vision, G. (1998) Perceptual content. *Philosophy* 73: 395–428.

CHAPTER 2

Towards an (improved) interdisciplinary investigation of demonstrative reference

Amanda Brovold and Rick Grush

I   INTRODUCTION

This chapter will have three substantive sections, all organized around an interdisciplinary understanding of demonstrative reference. Section 2 will focus on perception, and in particular perceptual mechanisms that allow us to latch onto entities in the environment as candidates for thought or reference. Our point here will be brief and, once made, we hope uncontroversial: current thinking notwithstanding, location is not the key to understanding how perceptual mechanisms single out their accusatives. Rather, it is gestalt criteria, or similar low-level principles, that do this. Given the way research is currently most commonly preformed, it is easy to see how this could be missed or ignored. But the focus on location is (at least in part) an artifact of choices about research. In any case, we can see that it is incorrect. Our conclusion will be: the basic perceptual accusatives, those entities, broadly understood, which are isolated and tracked by the perceptual system, are not physical objects, or even Spelke objects or 'proto-objects', or even locations in visual space. They are rather what we will call *gobjects* (for *gestalt object*), which is anything isolated by gestalt criteria. These often, *but not always*, involve space, hence the confusion. The focus on vision of most work in this area has aided this error.

In Section 3 we will turn from perceptual psychology to language, and in particular to an initial exploration of exophoric demonstrative reference. The goal of this section will also be straightforward: even a casual look at actual language makes clear that demonstrative reference felicitously tracks the full range of gobjects, including physical, spatial, Spelke, non-visual

We would like to thank the following people for various forms of help at various stages in the development of this project: audiences at San Raffaele University in Milan, Wissenschaftskolleg in Berlin, Simon Fraser University, the University of California, San Diego (UCSD), Cognitive Science Blending Group, the UCSD Department of Philosophy, Fey Perril, Maria Doulatova, Peter Yong, Adam Streed.

and non-spatial, and gestalt objects (gobjects). Linguistic demonstratives, like perceptual systems, are not beholden to space or location.

In Section 4 we turn from these brief preliminary points to a fuller discussion of demonstrative semantics, by way of outlining an account of the semantics of natural language demonstratives we develop elsewhere. The target of this account is the core semantic value of demonstrative expressions in natural language, both exophoric and endophoric. As will emerge, an adequate account of the semantics of demonstratives will interface only minimally with work on object perception, though there will be a great deal of interface with other kinds of empirical research.

2   PERCEPTION, VISION, SPACE, AND GESTALT PRINCIPLES

In this section we want to make two different but mutually supporting points. First, if what we are interested in is the basic principles that the perceptual system employs to isolate an entity and track it over time, then location is not the deep principle. Second, theorizing about perception has been influenced by an unbalanced diet of work specifically on vision, at the expense of research in other modalities and multimodal perception.

2.1   *Location vs. gobjects*

The view that spatial location is the key to understanding our ability to have meaningful perceptual contact with the world has a long history, but Campbell's (2002) work is the most influential recent example from the philosophical literature. Campbell, like Austen Clark (2000), has been impressed with the following line of thought: we are interested in perceptual objects, and at a minimum, such an object must be understood as something that is perceived as having properties, about which we could ask and answer questions such as 'Is it red?' They are potential truthmakers. To fulfill this role, sensory properties must be understood to be properties *of something*, something that can have more than one property (like shape and color). So we look to some relevant science to tell us what is involved in this binding problem, the binding of different sensed features together as features of a perceived *object*. One answer to this question has been that spatial location is the key. The visual system, to mention the result of one line of research, uses location in visual space at which properties are sensed to decide which properties to bind together as properties of a single object. Treisman's (e.g. 1996) work on this issue has had a significant influence.

There are two potential problems with this line of thought. First, the scientific studies by and large fail to distinguish location proper from binding effected by gestalt principles, because the relevant stimuli are examples of both. A red square at a location is both an example of a stimulus at a location and is also a degenerate example of a stimulus isolated by gestalt criteria, since the various parts of this region are similar, continuous, closed, proximal, and so forth. This is no knock on the scientific work, since for the most part the scientists aren't trying to distinguish location from gestalt organization, but to distinguish location from other things entirely.[1]

The mistake is natural. Perception latches onto entities isolated by gestalt principles. These principles presuppose a domain over which they are to be assessed. For example, the principles of proximity, continuity, and common fate presuppose some domain in which the raw sensory inputs are proximal, continuous, or through which they can move together. Quite often, especially in vision, this domain is space, and this has led many to mistake spatial location as the essential element of object perception. But as we shall demonstrate, *even when the objects of perception are spatial, and perceived through vision*, spatial location is not the theoretically deep principle. Entities can be isolated as objects of perception both when the object itself does not occupy a single bounded location (different locations, one object), and also in situations where an object spatially overlaps a distinct object (one location, different objects). Furthermore, in other modalities, such as audition, the ability of the perceptual system to isolate perceptual objects independently of spatial location has been empirically confirmed.

Even if we restrict attention to vision and entities that manifest themselves in space, location can be doubly dissociated from 'perceptual object'. First, to show that location is not necessary, see Figure 2.1.[2] Here the relevant stimulus does not span a continuous location: it is

---

[1] To take one example emblematic of the slide: in Calvert et al.'s (1998) review article on the cross-modal identification of objects, they comment that 'temporal and spatial proximity are clearly major determinate for co-registration' (248). Their citation for this claim is *A Sourcebook of Gestalt Psychology* (Ellis 1938). Much of the data discussed in their review in fact ignores the role of most of these criteria, because they focus on those few that are coextensive with location (contiguity and proximity). What begins as a reference to one gestalt criterion among others, proximity, ends up looking like an argument that that criterion alone is what is doing all the work. Further, this criterion itself is then oversimplified as location is treated as a proxy for it.

[2] One might be tempted to respond to Figure 2.1 by saying that the percept *appears* to be an object that covers or fills a contiguous location. This is right, of course. But the point is that this fact is the *result of* the operation of perceptual mechanisms that isolate trackable entities. The view we are engaging with is one that takes contiguous spatial location of the stimulus to be the *starting point* that determines what the perceptual system can take to be an object.

Figure 2.1 Partially occluded object.

spatially discontinuous yet seen as a single object. Figure 2.2 can be naturally seen as a three-by-three grid of objects on the strength of the gestalt principle of proximity, despite the fact that these objects consist of spatially separated parts. And in Figure 2.3, we have a perceptible object which stands out from the ground, not due to its location, nor its having boundaries (it does not), but to the similarity, continuity, and proximity of its parts, even though there are other elements of the stimulus that are as proximal to, but not perceived as part of, the object (the enclosed circles).

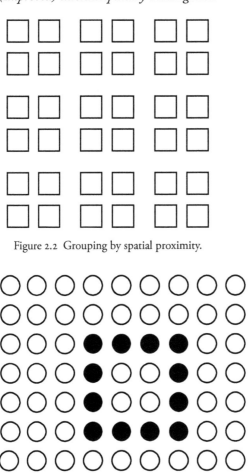

Figure 2.2 Grouping by spatial proximity.

Figure 2.3 Grouping by similarity.

Finally, Blaser et al. 2000 have experimentally studied situations in which what are perceptually presented as two distinct objects cover precisely coinciding spatial locations. The objects are Gabor patches, and they are a clear demonstration to the effect that 'being at one spatial location' is not sufficient for perceptually given material to be perceived as a single object (see Blaser et al. 2000 for details).

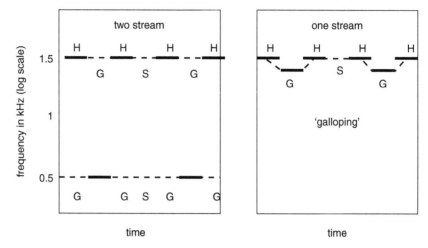

Figure 2.4 Grouping by non-spatial proximity (kHz, kilohertz; log, logarithmic).
Source: Bregman 1994.

The point is corroborated by research in non-visual modalities. Bregman (1994), for example, shows conclusively that gestalt organization is present in audition, and that it allows us to perceptually organize sonic stimuli into distinct trackable objects in situations where location is not involved in any relevant way. Among his points is to demonstrate how gestalt principles operate in non-spatial domains, such as pitch. A first example, illustrated in Figure 2.4, a sequence of three tones – high, low, high – is repeated. Depending on how proximal the low and high tones are, the stimulus is perceived either as a single repeating triple-tone sound object with a galloping cadence (if proximal), or as two separate objects, an intermittently beeping high tone and an intermittently beeping lower tone (when not proximal).

In the situation illustrated in Figure 2.5, subjects are presented with a stimulus that consists of eight overtones. Each maintains a constant pitch for the first 2 seconds. Then, three of the overtones rise and fall in pitch in unison four times before returning to their original constant tone. It is impossible not to hear these three overtones as a distinct auditory object emerging against the background of the remaining five. And this perceptual parsing is based on the principle of common fate, since the three overtones rise and fall together. These are only two of dozens of powerful examples illustrating how the perceptual system manages to parse the

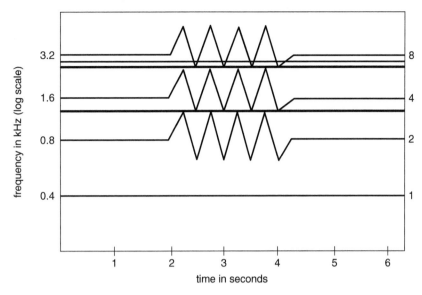

Figure 2.5 Grouping by non-spatial common fate (kHz, kilohertz; log, logarithmic).
Source: Bregman 1994.

sensory input into distinct trackable entities on the basis of gestalt principles, even when location is in no way involved in this parsing or tracking.

The perceptual system isolates these gobjects, assigns an object file to
them, binds features (e.g. pitch, volume, timbre), and tracks them over
time (e.g. one can tell that the sound that started at a low pitch ended at
a high pitch).

A certain kind of misunderstanding threatens here. Consider the following passage from Raftopoulos (2009):

... studies by Spelke (1990), Spelke et al. (1995) and Karmiloff-Smith (1992)
strongly support the assumption that the infant, almost from the beginning of
his life, is constrained by a number of domain-specific principles about material objects and some of their properties. As Karmiloff-Smith (1992, p. 15) points
out, these constraints involve 'attention biases toward particular inputs and a
certain number of principled predispositions constraining the computation of
those inputs'. Among these predispositions, Spelke asserts, are the conception of
object persistence, and four basic principles (boundness, cohesion, rigidity, and
no action at a distance).

The *cohesion principle* dictates that 'two surface points lie on the same object
only if the points are linked by a path of connected surface points'. This entails
that if some relative motion alters the adjacency relations among points at their

borders, the surfaces lie on distinct objects, and that 'all points on an object move on connected paths over space and time'. According to the *boundness principle* 'two surface points lie on distinct objects only if no path of connected surface points links them'. ... Finally the *rigidity* and no *action at a distance* principles specify that bodies move rigidly (unless the other mechanisms show that a seemingly unique body is, in fact, a set of two distinct bodies) and that they move independently of one another (unless the mechanisms show that two seemingly separate objects are in fact connected).

The force of these principles is such, that Gestalt principles are always overridden by the principles underlying the perception of objects in motion. These mechanisms allow, for instance, infants to infer that under proper movement a single object is displaced, despite the fact that this object is center-occluded by another superimposed object, a fact which could have led the infants to perceive two separated objects. (101–102)

There are two points to make about the relation between these remarks and those we intend to make in this section. First, a good chunk of the principles appealed to, and claimed to override gestalt principles, are themselves in fact gestalt principles. The movement of a single object displaced when center-occluded, for example, is an instance of common fate, since the two non-occluded ends of the object move together. The cohesion principle, as defined in the above quotation, follows from closure: the bounds of the set of points connected by other surface points defines a closed boundary. So the first point is that many of these principles are, or are reflections of, gestalt organization, not competitors to it.

The second and deeper point concerns being clear on what it is we are trying to explain. Our topic is the basic principles that determine what the perceptual system isolates and tracks. Of course most higher nervous systems have additional abilities. For example, humans, even human infants, have knowledge about physical objects and constraints they are subject to, that go beyond what the basic perceptual system is keyed into. The result is that it is possible for a human subject's perceptual system to isolate something, track it over time, and collect information about it over time in an object file, such that the subject realizes that this tracked entity cannot be a physical object. It seemed to pass through a solid barrier, for example.

We should not confuse those principles that allow a subject, even an infant, to express surprise at something behaving in a certain way, perhaps because that behavior is inconsistent with how it assumed objects of the presumed type behave, from principles that allow the perceptual system to isolate and track entities in the basic, broadest sense. Many of the sorts of studies in developmental psychology appealed to are aimed at the infant's understanding of material objects or bodies or causality, and so forth. This is well and good. But it is a mistake to assume that what we learn from the study of the infant's capacities are the principles that the *perceptual system* uses to isolate and track entities to begin with. And expressions such as 'the infant … is constrained by a number of … principles', as in the above quotation, invite such slides by being vague as to what aspect of the infant's competence is at issue – the infant's ability to perceive, the infant's ability to know what counts as a material body, and so forth. Indeed, the perceptual system's ability to isolate and track gobjects is a precondition for the application of these more specialized categorizations. For example, the infant's surprise when viewing something that apparently travelled through a barrier is *precisely* surprise at seeing *a single perceptual accusative* – something that the perceptual system isolated and tracked as a single entity before, during, and after its passage through the barrier – do something that material bodies aren't supposed to be able to do. If the perceptual system weren't tracking the gobject and collecting information to the effect that it was violating this or that physical object constraint, there would be nothing to be surprised about.

To summarize so far, in the case of the Gabor-patch gobjects (from Blaser et al. 2000), we have two distinct trackable entities even though there is no difference in their spatial location. In the cases of continuity and proximity (see Figures 2.1–2.3), we have entities bound as trackable gobjects despite the fact that they are not in a contiguous location. In scientific parlance this is a double dissociation. In philosophical terminology we can say that location is neither necessary nor sufficient. And while in these examples space is still present – it is the domain over which gestalt principles are assessed – the auditory examples have no relevant location aspects at all.

2.2 *Vision vs. multimodal perception*

Over-reliance on vision research not only can mislead about the role of location in perception, it also downplays the fact that much of actual perception is from modalities other than vision, and is also often multimodal. Many researchers slide from 'perception' to 'vision' with such ease that the slide is not even noticed. A failure to appreciate this invites the danger that what are contingent features of one perceptual modality will be mistaken for essential features of perception. For example, Scholl (2007) begins his paper by explicitly noting that '[p]sychological research on object persistence is not isolated to visual perception' (564), yet by the end of the first section he has moved on to focus on what he takes to be the core mechanisms for the phenomenon of object permanence. To be fair, Scholl does offer some justification of this move. He points out that the similarity in approaches to research may 'reflect an underlying connection: psychological mechanisms of object persistence (especially relevant parts of mid-level visual object processing) may serve to underlie the intuitions about persistence that fuel metaphysical theories' (564). This certainly could be true. It could also be the case, though, that similarity in research methodology is in fact preventing us from seeing the actual fundamental mechanisms underlying our abilities to perceive, track, and refer to objects. And indeed, when research manages to break out of this narrow methodology, we do see evidence for this.

The intersensory redundancy hypothesis, for example, suggests that objects about which we get information from multiple modalities are more salient than those we get information about from a single modality (see Bahrick and Lickliter 2000). This effect is not simply additive; the point is not the obvious one that more information is better, but rather that information through multiple channels is better. Though this effect is at times described using spatio-temporal coincidence as the default criterion for binding, the research suggests that this criterion is not in fact decisive. Bahrick and Lickliter note, for example, that asynchronous stimuli can be paired in some cases, and postulate that the sort of pairing necessary may be dependent on the modalities in question, as well as the particular stimuli.[3] A powerful example of the ability to pair multimodal stimuli comes from the ventriloquism effect, which has been shown not to be dependent on location in the typical auditory–visual pairings or in other modalities (see Keetels and Vroomen 2008). This effect is embarrassing for any account that takes it that apparent location is the principle used to direct binding, since it vividly shows that in fact it is binding (effected via the gestalt principle of common fate, specifically the common temporal profile of alterations in the visual stimulus of the moving mouth and the auditory stimulus of the speech) that is determining apparent location (of the auditory stimulus).

The particularities of how research is designed in fact make a tremendous difference in what the underlying mechanisms might appear to be. Of course there are legitimate reasons for choosing simple methodologies in research. But it is a mistake then to assume that 'by choosing simple

---

[3] They in fact do not explicitly note the impact of a particular stimulus, but their examples make it clear that this is built into the analysis.

nonfigurative objects and a simple task, we [are] able to find basic characteristics underlying object reference'.[4] And in fact as we have tried to illustrate, evidence to the contrary is not far to seek as soon as one decides to
look. Occluded pineapples, Gabor patches, and Bregman's examples (all
discussed in Section 2.1) clearly indicate that it is gestalt criteria that are
doing the work. Spatial location is only a special case.

2.3   Location vs. objects

Much of the scientific research in visual perception addresses the issue of
the relative priority of location and objects. Treisman's work, for example,
has argued that, from a neural systems standpoint, locations are prior to
objects. Other work, though, suggests the opposite, that the perceptual
systems track objects independently of location.

One productive research paradigm for investigating the identification
and tracking of objects is object-specific priming benefits (OSPBs). A typical OSPB set-up will have subjects fixate on a visual array that includes
two objects. The objects will then each have some symbol flashed on them
(priming phase). After the symbol disappears, the objects move to a new
location on the screen and a target appears on one of the objects. The
task is to judge whether the target was one of the primed symbols or not.
Though the task does show a general priming benefit (in that subjects are
quicker to respond when the target does match a primed symbol than
when it does not, regardless of location), the effect of interest is the difference between what are called congruent and incongruent trials. In congruent trials, the matching target is on the 'same object' as it originally
appeared on. In such trials, subjects are even quicker to respond. OSPBs
have been taken to be evidence for object files since the effect shows
that there is a priming benefit greater than a generalized priming benefit expected from any prior exposure to a target stimulus if that stimulus
reappears on the *same object* that it was primed on. The crucial bit is that
the effect follows the object *even when that object moves to the location at
which the other non-primed object originally appeared.* In these cases the
second letter is flashed at the same location, but on a different object, and
priming follows the object, not location.[5]

---

[4] Beun and Cremers 1998, 84. This sentiment is not unusual. Not many authors bother to express it
so clearly and explicitly, but it is clearly an implicit presupposition of much of the literature in the
relevant domains.
[5] Note that one can't try to save the location account by supplementing it with time, as though it is
spatio-temporal contiguity rather than merely spatial contiguity. For when it is asked which spatial

Interestingly, OSPB research seems to include some cases where the judgments of continuity of an object are made without that object obeying laws of continuity of movement or stability of location that are often taken as fundamental to our ability to distinguish and track objects. For example, Henderson and Anes' (1994) experimental set-ups include the priming and target objects appearing in the same location throughout the trial, but with the switch from prime to target occurring during a saccade in which the subjects move from a fixation point farther away from the objects, to a closer one. It is perhaps not surprising that this causes no problem in object tracking, as in everyday life we move our eyes to look to objects which stay in the same location. However, Henderson and Anes then follow that experiment with one they describe as mimicking the retinal events of the previous experiment, without eye movement. In this set-up the subjects maintain a single fixation but the objects move in from the periphery (during preview) to near the fixation (during the target display). As the authors note, spatio-temporal continuity is, in a literal sense, violated by this set-up since the objects jump to a new location. However, the manner in which they jump mimics the image subjects would receive if the objects stayed put while their eyes moved to focus on the objects. Evidently, in either of these scenarios we have no difficulty tracking the objects, as they show very similar OSPBs.[6]

OSPB research is primarily restricted to investigating the priming benefits from *visually* tracking objects. Even when typical visual objects are used, though, some such research has shown that our ability to identify and track objects is more flexible and (though the researchers do not put it this way) dependent on gestalt criteria than we might expect. For example, Gao and Scholl (2009) have shown that this effect is robust even

regions are the ones continuous with prior regions, there is no coherent answer other than 'those regions occupied by the object'. Unless there is an object being tracked as it moves, there is no principled way to single out any of the infinite number of spatio-temporally continuous traces as special.

[6] Focus on spatio-temporal continuity sometimes leads to the denial that an object appearing in a new location after disappearing from a different one (without following a continuous path) is interpretable as the same object. Scholl (2007), for example, says that '[i]f an object disappears at one location, and an object immediately appears at a spatially separated location, then those two instances cannot be the same object' (566). It isn't clear what this is supposed to mean. As a report on normal adult humans' theories of how physical objects in fact behave, it is possibly correct. However, it seems not to be a necessary feature either of our conceptual scheme (as many examples from science-fiction films attest), or of basic perceptual mechanisms. Apparent motion is an example where low-level processes treat as one object two spatially non-continuous stimuli; and at a higher level, people playing video games, such as Asteroids, have no problem seeing the ship or rock that disappeared on one side of the screen and instantly reappear on the other as the same object.

when objects are identifiable only by common fate. In other words, when the objects lack boundaries but instead consist only of groups of separated pixels moving together.

Further, one recent study (Jordan et al. 2010) has shown that OSPBs occur even when multiple modalities are involved. In the experiment, the basic set of classic OSPB experiments is for the most part preserved, except that the preview and the target cues are given in different modalities. Subjects are shown a visual array, and two objects (icons that appear briefly within two outlined areas) are previewed visually within that array. After the icons are no longer visible, the objects move (the outlined areas appear to rotate across the visual array). After rotation a sound is played from near one of the new locations that may or may not match with one of the icons previewed (for example, if the icon was of a phone, the sound may be a ring). The results of the experiment show effects along just the lines of single-modality OSPBs. That is, when the sounds are congruent with the post-movement location of the objects (which are no longer visually available – their icons are gone, only their outlines remain) the subjects are quicker to identify the sounds as matching one of the preview objects than when the sounds are in the incongruent location. Though location does appear to be playing an important role in this set-up that should not distract from the general lesson that multimodal research is important in this area. Combined with other evidence (such as the ability to bind multimodal stimuli without co-location as noted above) it is clear that spatial location here may just be a proxy for deeper criteria that are in fact doing all the work.

2.4 Discussion

Our point in this section is that the ground level at which perception isolates and tracks entities is one below spatial location, material bodies, Spelke objects, or what have you. This shouldn't come as a surprise. Our perceptual systems are a small refinement on a vast evolutionary history of simpler nervous systems that didn't have the wherewithal to reason about material bodies, and may not have had vision. But basic principles of the sort embodied in gestalt criteria are, computationally, relatively cheap means of latching on to things that more often than not turn out to be material or biological objects. When you track a shadow across a field, or a spotlight across a stage, you are not engaging a host of high-level mechanisms – such as an ability to perceive material bodies, together with doing metaphorical mappings – that then allow your perceptual system to isolate the shadow or light on analogy with a moving body. This might be how a twentieth-century analytic philosopher would design the system, but so much the worse for such a design strategy. Rather, you are simply engaging perceptual mechanisms largely similar to those that have the house cat chasing laser dots, fish fleeing fast-moving shadows, and bees tracking the position of the sun.

It might be objected that though perceptual systems can track non-physical or non-spatial gobjects, the fact of the matter is that what we are interested in is something like typical, or ecologically valid, cases, and how perception deals with them. Indeed, analysis of our ability to refer to and track objects generally restricts itself to 'everyday ordinary objects' either implicitly or explicitly. Take two fairly representative examples: Lowe (2005) lists the objects that philosophers most usually discuss and Scholl (2007) notes that his investigations apply only to typical everyday objects. Both list exemplars to give the idea of the class they mean to refer to: cars, tables, basketballs, cats, books, etc. While it may be reasonable to restrict the domain of objects an investigation is responsive to, it is also important to be cognizant of and realistic about the nature of that restriction. (Compare: a biologist might claim that trout and bass are typical fish, and thus it is reasonable to restrict our study to such cases, and furthermore that what we learn from them will apply to all fish.) It is certainly true that the objects in the category are fairly uncontroversial exemplars, but it is not the case that all entities not in that category are 'unnatural' objects, creations of academic fantasy that can be safely ignored.[7] The tracking of shadows or individual instruments in a symphony may not be the prototype of seeing a physical object, but they are hardly freakish creations of ecologically invalid science or irrelevancies of armchair philosophy.

It is worth remembering the point of the enterprise, which was to use scientific investigation of how perception isolates and tracks objects *to inform* philosophical, semantic, and epistemological investigations into our ability to think about and refer to entities in the world. If empirical investigations, such as those of gestalt psychology in general, or Bregman's (1994) or Blaser et al.'s (2000) work in particular, suggest that location is not essential to isolating and tracking objects we can perceive and talk and think about, then that is what science says, and we need to adjust our

---

[7] Scholl suggests this is so when he notes that he does not mean to consider such things. He also, a bit mysteriously, excludes objects that 'spatially extend beyond our immediate experience' (565n1). Though Scholl acknowledges at this point that the question of what counts as an object is a significant one, he takes it to be irrelevant to his research. However, it is certainly relevant to deciding the acceptable reach of his conclusions.

high-level philosophical theories and expectations accordingly. To limit which scientific results or protocols we take account of by reflection on what kind of objects or modalities we think are important or interesting for philosophical, semantic, or epistemological purposes is to do things the other way around. While we can't say such an approach is objectionable *tout court*, it certainly loses its claim to being empirically informed or responsible.

3   EXOPHORIC DEMONSTRATIVE REFERENCE AND GOBJECTS

One possible response to the previous subsection would be to accept that at the ground level perceptual systems key in on gobjects, but then claim that since our overriding concern is the interface of perception and linguistic reference, we need to focus our attention not on this ground level, but at some slightly higher level, those gobjects that are genuine objects of public discourse and reference, entities about which we can discourse, correct each other, act on. Maybe these will be those entities that satisfy the constraints on material bodies that the developmental psychologists investigate, for example. And if this is right, then the way to conceive of the interdisciplinary investigation is as teamwork between semantics of demonstrative expressions and something slightly higher level than low-level perception, perhaps developmental psychology geared towards infants' ideas about material bodies or what have you. The problem with this line of response is that when it is explored, it becomes evident that the linguistic phenomena are at clear odds with the premise. Gobjects – even non-physical, non-spatial gobjects – are perfectly felicitous objects of demonstrative reference, even supporting subsequent pronominal coreference.

Consider some examples:

1. *This* rock is too heavy. I can't move *it*.
2. Look at *that* shadow moving across the field. *It* keeps going in circles.
3. *That* shadow stretches across the entire field. A single beam of light falls across *it* around midfield.
4. What is making *that* noise? I can't concentrate with *it* blasting like that.
5. *That* silence creeped me out. Didn't you notice *it*?
6. *This* uncomfortable silence lasted the entire trip. *It* was broken only by the occasional request for bathroom breaks.

In (1) we have a prototypical demonstrative reference to a physical object. In order to prevent an objection to the effect that these demonstratives aren't really referring, we have a coreferential pronoun in the following sentence – where there is coreference there is surely reference. As (2) shows, such reference is not limited to physical objects, but can be along the lines of a Spelke object, a coherent bounded region. But as (3) shows, this region's boundaries can be determined by gestalt criteria, which can override strictly region-based criteria. Note that there is no sense in which the shadow exists unperceived behind an occluding beam of light. It is construed as a single entity occluded by a light on the basis of gestalt criteria. Even percepts without any relevant locational features can be referred to, as in (4). Note that the demonstrative referent is clearly the noise, not the thing making the noise – it must be, since this is part of a description used by the full NP (noun phrase) to refer to the thing making the noise. And as (5) demonstrates, even a continuous absence of sensory stimulation can be bound, tracked, and referred to. It can even be bound across temporal discontinuities, as in (6).

But what about extended discourse, tracking, and so forth? Consider the dialogues in (7) and (8):

7. A. Do you hear that?
   B. Just barely. It's very quiet. Can't even tell where it's coming from.
   A. Is your stereo broken?
   B. No, my stereo isn't making that noise.
   A. Thunder? Maybe there's a storm approaching?
   B. Doesn't look like it.
   A. Chainsaw?
   B. No, chainsaws don't beat like that.
   A. Oh, it's a helicopter!

In this exchange, the interlocutors are trying to identify the source of a sound that they hear. Their perceptual systems have isolated an auditory percept, and are collecting information about it. They can then refer to it with demonstratives, and coreferential pronouns. They then go on to discuss its features, ask and answer questions about it, and so forth. All of this machinery is working perfectly well in absence of any relevant locational information, and in absence of any identification of its causal source. That there are some situations where the auditory stimulus we isolate and track is then recognized as a sound probably produced by something we could also perceive visually, such as a helicopter, is an interesting phenomenon. But it is one that goes quite beyond the present issue, which is

understanding how perceptual mechanisms isolate and track their accusatives, and the equally plain fact that we can refer to these accusatives with
demonstratives. And in fact, such higher-level achievements – such as recognizing a sound as one produced by a certain kind of artifact – *depend
upon the fact that one is able to isolate the sound on purely auditory grounds
to begin with*, in order to track and then recognize it.

One might feel an intuitive pull that anything we identify as an object
must be located in space because in order for us to identify something as
a public, objective entity we must be able to act on it in some way. Since
action is typically bodily, and we act on things by comporting ourselves
with respect to them in space, this might seem to suggest that objects must
be spatial in some sense. It is not obvious to us that action on an object
needs to be a requirement on thought, in fact it smacks of a desperation
Hail Mary to preserve a central place for space in the theory.[8] But even
if it is accepted as a requirement, it does not help. Consider the following fictitious, but entirely natural exchange between two sound engineers
working on a mixing board. In this example, the electronic composition
being worked on is not stereo, and so there is no apparent spatial or location difference in any of the sounds the engineers are referring to:

8. A. I think the pure tone fill is too loud.
   B. Which one?
   A. (*Manipulates a lever making volume of one of the dozen components
      of the sonic landscape fluctuate up and down for a second or two.*) This
      one.
   B. Ahh. OK, let's hear it again.
   A. What do you think?
   B. Yeah, it seems too forceful. Maybe keep the same volume, and just
      take it down an octave?
   A. Sure. How's that?
   B. Much better.

In this electronic composition, the sounds are not even intended to
resemble those produced by physical instruments, and so there is no issue
of calling them 'that violin', nor any issue of thinking that what is being
talked about is their causal source. The different elements in the composition are isolated and identified purely on their non-spatial audible

---

[8] It might be that what is driving this feeling, though, is a sense that demonstratively referring to an
object requires some connection with it. There is *something* to this idea, but clinging to the possibility of bodily interaction misses the point. See the discussion of various sorts of control in Section 4.

features: volumes, tones, timbres, and gestalt-principle grouping effects as discussed by Bregman (1994). But this is more than enough. The interlocutors can identify sounds, discuss their properties, and act on them to bring them to the other's attention or to change their features. And though the agent acts on them in a spatial manner (she moves a lever, for example), what she is acting on need not be individuated spatially. (It is not difficult to imagine, with technology available today, sound engineers who through neurofeedback training could manipulate sounds purely by willing them to change in various ways, and the connection to space would be entirely severed with no obvious detriment to what it is theoretically important to investigate.)

It will be helpful to summarize Sections 2 and 3 before proceeding. Recall, the phenomena we are interested in are (i) our perceptual contact with entities we think about and refer to, and (ii) how this contact allows us to think and communicate about the world. And the conclusion of these two sections is clear: both the perceptual mechanisms themselves, as well as the capacity for the formulation of thoughts, and communication in public language using (exophoric) demonstrative expressions are keyed into gobjects. It is true that for reasons that go beyond these basic facts we often are interested in certain subsets of these accusatives, such as physical objects that conform to more restrictive constraints than those the perceptual system itself imposes, or entities that can be perceived through more than one modality. But whatever considerations might make such subsets more interesting for certain purposes are tangential to the phenomenon itself. And importing features of these restricted cases back as though they were essential features of the basic phenomenon of interest is, while a *natural* mistake, a mistake.[9]

4   TOWARDS AN INTERDISCIPLINARY STUDY OF DEMONSTRATIVE REFERENCE

Sections 2 and 3 we addressed the issue of whether spatial location was in any way essentially linked either to perceptual processes that isolate and

---

[9] It has been suggested to us that talk of, e.g. silences and shadows is metaphorical, with physical objects as the metaphorical source domain. We see no reason to take them as metaphors. Is there really any plausibility to the suggestion that the toddler who sees – using the same evolutionarily ancient perceptual tracking mechanisms available to a crawdad – a shadow sweep across a field and ask 'What is that?' is constructing a metaphor whose target domain is physical objects? The idea that they are seems motivated by a prior commitment to physical objects being the primary accusatives of perception. But the question would be easy to resolve by empirical investigation. And we have no hesitation as to the side on which we are placing our bets.

track objects, or reference to such objects via demonstrative expressions in natural language. The result was that they appear not to be connected in either way.

This is not to deny the value of an interdisciplinary investigation of the semantics of natural language demonstratives. It just means that it will be of a different character than is typically assumed. In this section we briefly outline an account of demonstrative semantics that we more fully develop elsewhere (Grush and Brovold, in preparation). As will be seen, it is a very interdisciplinary proposal, though the specific sort of interdisciplinary synergies it appeals to are of a quite different character than is typically the case.

4.1 The control profile

We must begin with some considerations that might initially seem tangential. Semantic theory (in linguistics at any rate, not so much in philosophy) has over the last couple of decades increasingly recognized the centrality of conceptual archetypes in understanding the semantics of natural language expressions.[10] We claim that the conceptual archetype that lies at the heart of demonstrative semantics is what Langacker (2002) has called the *control cycle*. A schematic description of this cycle is as follows: in the initial *baseline* phase an agent has control of some set of entities; in the second *potential* stage some new entity emerges as a candidate for control, the agent notices it; in the third *action* stage, the agent acts so as to bring the entity under its control; in the final *result* stage, the entity is in the agent's control; in a potential fifth *loss* stage (not part of Langacker's formulation, but arguably part of the archetype), the agent may lose or relinquish control of the entity.

The prototype of this cycle is physical control, of course. But the control cycle can also manifest itself in other domains, including perceptual, attentional, epistemic, and social. One (*but only one!*) linguistic manifestation of this archetype is the pervasive use of lexical items for various phases of this cycle regardless of domain. Verbs like *get*, *have*, *lose*, are all prototypically geared towards aspects of the physical control cycle, but apply equally felicitously to perceptual, attentional, and social control:

---

[10] For a few historically influential representative examples, see Talmy's discussion of *force dynamics* (1988); Lakoff's famous discussion (following the doctoral dissertation work of Claudia Brugman) of the English *over* (1987); and for applications of conceptual archetypes to clause-level argument structure, see e.g. Goldberg 1995.

9. A. Do you have the wrench?
   B. Yeah, I've got it. No, wait, I lost it. [Physical control]
10. A. (*Sniper team spotter to sniper*) Do you have the guy on the west tower?
    B. Yeah, I've got him. Wait, I lost him. [Perceptual control]
11. A. Do you have another car?
    B. Yeah, I got one on my birthday, but I lost it in the divorce. [Social control/ownership]
12. A. Did you have something you were going to tell me?
    B. Wait, I had it just a second ago, but I lost it. [Attentional control]

These parallelisms among a handful of lexical items represent only a tiny fraction of evidence for the centrality of this archetype in influencing grammar and lexical semantics. The interested reader should see Langacker (2002) for details. We would also like to point out that, while we use Langacker's terminology because we find his analyses powerful and revealing, our account does not depend on the details of Langacker's proposal.

Furthermore, if one dislikes the idea that all these different domains can be described as different kinds of control, then just use a another word in your mental-reading voice as appropriate. Our analysis doesn't depend on this. All we appeal to are the surely uncontroversial claims that (i) people can seek, acquire, and lose physical, perceptual, social, and attentional control of entities; and (ii) people in close proximity, e.g. when engaged in a conversation, often have a good idea of what their own and their interlocutors' control relations (physical, perceptual, etc.) are to various entities; and (iii) speakers often have a good idea of the proclivities for control of various sorts. Knowledge of the sort described in (i)–(iii) is what we will call the *control profile* of a speech situation at a time. It is the rich, shared contextual knowledge interlocutors have of their own and of each other's control relations, and each other's knowledge of each other's control relations, to potential referents, especially as they evolve during the course of a conversation.[11] Just to be very clear about one point: the key issue is not one of actual, successful control. *The control profile is shared knowledge of*

---

[11] One crucial element to how the control profile evolves during a conversation is that the use of demonstratives, which *relies on* the control profile, also *changes* it. The use of a sentence such as 'Look at that cup' not only exploits the control profile as it is at the moment before the sentence is uttered (as we shall explain), but it immediately alters the control relations, since the addressee will be attending to the cup after hearing the sentence in a way that differs from how she was attending to it before hearing the sentence. Few semantic categories have this feature – characterizing something as 'red' or 'above' doesn't change its color or location. This is one of the phenomena that have made understanding demonstrative semantics so difficult. We shan't explore this here, but see Grush and Brovold (in preparation).

all aspects of the control cycle, including: which entities your interlocutor is not paying attention to, is less likely to be able to perceive or grasp, what they used to, but no longer, own. Depending on context and language, any of these factors can licence or influence demonstrative use.

4.2 *Proximal and distal in exophoric anaphora, in physical-control situations*

Though our goal is to give an account of the semantics of demonstrative expressions as a semantic category, it will be convenient to begin with one element of demonstrative semantics, the so-called proximal/distal contrast, and show how aspects of the control profile do a better job of explaining the factors involved in this contrast than competing accounts.

All languages have demonstratives, and most have at least a proximal/distal contrast, exemplified in English with *this/these* (prox) and *that/those* (dist). Diessel (1999) gives an admirably comprehensive survey of demonstratives in a variety of languages, for example, and he makes great use of 'proximal' and 'distal' labels for the most basic contrastive forms found in most languages.[12] The distinction between them, as their labels suggest, is often explained in spatial terms: proximals are preferred when the referent is near the speaker, distals when it is not near the speaker. It is certainly true that to a first rough approximation, patterns of English 'this' and 'that' correlate with distance, and in fact empirical studies confirm this (see below).

However, as always, we need to be alert to the possibility that this could be because both (a) patterns of usage and (b) distance to referent correlate with some third factor. Because we typically are in control, or are better able to take control (physical, perceptual, and attentional), of nearby objects than distal ones, it is possible that the contrast is really tracking actual or possible control relations – elements of the control profile. And in fact a variety of considerations confirm that it is some form of actual or potential control, rather than spatial distance, that is the crucial factor.

---

[12] It is notable also that demonstratives in other languages often also carry a variety of other features. This is a point we will return to later when we discuss core demonstrative semantics. In his cross-linguisitic survey of demonstratives, though, Diessel points out that the idea that demonstratives are essentially spatial deictics is a significant simplification. While most languages include demonstratives that mark distance in some manner, there are some which include 'distance-neutral' demonstratives (38). Moreover, when demonstratives do (seem to; see below) code for location, there is great variability in the way that this is done.

Evidence for this can be found by examining situations in which control and spatial proximity come apart. For example, Italians all learn in school that *questo* (prox) is for things closer to the speaker, while *quello* (dist) is for things farther from the speaker (or near a third party), and *cadesto* for entities near the addressee. However (as in English), it is perfectly felicitous to ask *Che cosa è questo?* (What is this?) of a small wound that you see on someone's back, perhaps something the addressee might not know about.[13] In the prototypical case, the speaker might touch the wound, and perhaps the addressee is not initially aware of it. But being part of the addressee's body, it is hard to see that the referent is not closer to the addressee. Of course, if the speaker touches it, then it might be argued to be closer to the speaker. But note that control is the operative notion: if the speaker has a 2-meter stick that she uses to touch the wound, she is relatively spatially distant, but the stick extends what she can control and influence. Regardless of distance, if the speaker has the means to control the wound (by touching it), then *this* is felicitous; however if the speaker does not – if the speaker is standing a few meters away, referring to a wound on the addressee's hand – then *this* is infelicitous, and *that* is preferred.

Empirical studies confirm this. Coventry et al. (2008) showed that in both English and Spanish, while the use of proximals vs. distals correlated with spatial distance when distance was the only independent variable, subjects who were given tools that extended the range of what they could influence (a manipulation that did not affect the spatial distance) had a concomitant extension in the range of referents for which they used the proximal.

4.3   Proximal and distal in exophoric anaphora, in perceptual and social-control situations

But physical control is only one aspect of the control profile. Perceptual, attentional, and social control also influence proximal vs. distal felicity, though often physical control has a dominant influence. The general idea is that proximals in English (and some other languages) prefer entities over which the speaker has control (and often but not always over which the speaker is giving the listener control), and distals for things over which

---

[13] Given the definitions above, *cadesto* would seem preferred; however, it is not. In part this is because *cadesto* is falling into disuse, except in parts of Tuscany. But given the usual definitions, either should be better than *questo*, as the referent is physically far from the speaker and close to the addressee. Nevertheless, as in English, *questo* is perfectly felicitous, even preferred.

the speaker does not have control, or is losing or relinquishing control. Some languages also have demonstratives that center on the addressee's control, as in Italian *cadesto*.

Consider a potential referent in a box that the addressee is clearly looking at with interest, and which the speaker, though no farther from the box, is unable to see, and the speaker is clearly not endeavoring to move such as to be able to view the referent. In such a case, the *addressee* has perceptual control over the referent (and the speaker knows this), but the speaker does not. The speaker can felicitously ask (13), but (14) is either bad, or questionable, or requires a special context:

13. What's that?
14. *What's this?[14]

However, if the roles are reversed – the speaker is the only one who can see the referent, and knows that the addressee can't – then the pattern reverses:

15. Look at this.
16. ?Look at that.

A striking example of the influence of social control was demonstrated by Coventry et al. (2008), who noted that manipulating who last touched an object (a subtle but powerful indicator of ownership/social control) influenced whether speakers preferred proximals or distals. Subjects were less likely to use a proximal for referents when they saw the referent placed at its location by another agent, when all other factors are controlled (e.g. relative distances between the referent and the conversation participants at the time of the utterance). Coventry et al. merely noted the effect but did not provide a theoretical explanation, and so they should not be interpreted as endorsing our interpretation. The influence of perceptual and social control on demonstrative use is far more complex than these brief comments. (See Grush and Brovold, in preparation, for more detail.)

4.4 *Endophoric anaphora*

Exophoric demonstrative anaphora, where the referents are entities external to the discourse, contrasts with endophoric demonstrative anaphora,

---

[14] Though note that if we change the context and provide the speaker with physical control, even when lacking perceptual control – for example by having the box in the speaker's hands, perhaps showing its open end to the addressee so that the addressee can inform the speaker about what is inside – then *What's this?* becomes felicitous again.

where the referents are internal to the discourse. This includes both reference to entities identifiable only through the discourse (*When I got on the bus today there was* this *guy singing at the top of his lungs. Turns out that guy was a cop.*), and also elements of the discourse, such as propositions, themselves (*John was late again.* That's *why nobody invites him to surprise parties.*). In these situations physical and perceptual control are typically not at issue. Rather, attentional and epistemic control are at issue.

For example, propositions that have just been mentioned in previous discourse are under both the speaker's and addressee's attentional and epistemic control, while those that the speaker is about to mention, but hasn't yet, are under the speaker's attentional control but not the addressee's. In fact, it is striking to note the parallel patterns of proximal and distal use in physical, perceptual, and attentional (endophoric) control scenarios, as shown in (17)–(19).

In (17a) and (17b), the addressee has just eaten a cookie, and the speaker has a different cookie on a plate in his hand he is offering to the addressee. The speaker has physical control of the second cookie and is endeavoring to put it in the addressee's physical control. *That* is being used to refer to the just-eaten cookie, and *this* to the one being offered:

17a. If you thought that cookie was good, then try this one.
17b. *If you thought this cookie was good, then try that one.[15]

In (18a) and (18b), the speaker and addressee are sitting side by side while between them are two vertically stacked computer screens. The addressee is examining an image on the top screen, and the speaker an image on the bottom screen. The addressee says that the top image is overexposed. The speaker then wants to move the addressee's attentional control from the top image to the bottom image where her (the speaker's) attention is focused:

18a. If you think that image is overexposed, take a look at this one.
18b. *If you think this image is overexposed, take a look at that one.

In (19a) and (19b), the speaker has just told the addressee a joke, which the addressee claimed was childish. The speaker then intends to tell the addressee another joke, that is, to move the addressee's attentional control from the previous joke to the yet-to-be-told one:

---

[15] Of course if the control profile is altered (17b) can be made felicitous, for example if the speaker is holding half of the just-eaten cookie, and the uneaten one is on a table out of reach. The felicity assessments made are on the basis of the example as described in the text, though.

19a. If you thought that was a childish joke, get a load of this.
19b. *If you thought this was a childish joke, get a load of that.

The common thread here is that *that* is preferred for the entity over which both speaker and addressee have had control, but over which the speaker wants to relinquish control in order to give the addressee control of something else – a new entity, referred to with *this*, over which the speaker already has control. This is not just a pattern in English. For example, a language of New Guinea, Ambulas, includes what Diessel (1999) labels 'manner demonstratives' which are discourse deictics. Manner demonstratives have three forms, as do other demonstratives in this language, and Diessel labels them as 'proximal', 'medial', and 'distal' according to their morphological connection to the other demonstratives of the language. The way that these discourse deictics operate, though, has nothing to do with physical space, or indeed even distance in temporal or discourse space. Instead the 'distal' form is used to refer back to previous elements of the discourse while 'medial' and 'proximal' forms refer to information yet to be introduced by the speaker (Diessel 1999, 17).

4.5   *Control and core demonstrative semantics*

As we have illustrated our proposal so far, there is a fairly obvious worry. The worry is that there are demonstratives in many languages that are keyed into spatial features more specific than proximal and distal, and it might not be at all obvious how an account based on the control profile can accommodate them. For example, several languages include demonstrative forms that differ according to the elevation of the referent (see Diessel 1999, 42). For example, Khasi, spoken in the Himalayan area, includes demonstrative forms for the common 'proximal', 'medial', and 'distal' distinction as well as for elevations.[16] Given the significant number of such demonstratives in many languages, and given that in such cases spatial considerations cannot be explained away as correlates of aspects of the control profile, perhaps the idea that the proximal/distal contrast is not primarily spatial should be resisted.

Our response is that what we intend to provide an account of is what we call *core demonstrative semantics*. This is the set of semantic features whose presence is something like necessary and sufficient for an expression

---

[16] Khasi, like many languages, also includes a form that is used for referents that are not visible. This is an embarrassment for theories that draw a tight link between visual object perception and demonstrative semantics.

to be a demonstrative (or to be used demonstratively; see below). That some demonstrative expressions include semantic specifications beyond this core is interesting, but not part of our concern.

For example, the Italian counterparts of English *this* and *that*, *questo* and *quello*, have a gender specification. Even in English, it has been noted that there are so-called 'demonstrative uses' of personal pronouns (Kaplan 1989a and 1989b), for example pointing at someone in a line-up and saying 'It was *her*.' In Khasi, it is true that the demonstrative root *-tey* has a spatial specification, higher elevation, of the sort that could easily be coded by a spatial preposition or other non-demonstrative locative (like English 'above'). But some questions a semantic theorist should ask are: What is the difference between English *her* used as a pronoun, and used demonstratively? What is it about the use of *-tey* that makes this expression a *demonstrative rather than a preposition or adjective* meaning *above* or *high*? What does English 'above' lack such that it is not a demonstrative with a meaning similar to *-tey*?

These are obviously important questions for a semantic theorist to ask, and in so asking, we claim that what the semantic theorist is asking after is precisely what we have called core demonstrative semantics. If this is right, then one cannot cast doubt on a theory of core demonstrative semantics by pointing out that this or that language has an expression that has some specifications over and above those detailed in the proposal. Rather, what would have to be shown is either that some expression which is pretheoretically clearly being used as a demonstrative lacks the proposed core semantic value, or that some expression which is pretheoretically obviously not a demonstrative has this core semantic value.

What is this core demonstrative value, then? The proposal is this: the core demonstrative semantic value is the exploitation of some aspect(s) of the control profile to identify the intended referent.[17] The control profile is a very rich thing, and languages have many choices about which elements of it to exploit, and how to exploit them. Many languages, including English, are fairly minimal, having only a contrast between *in control* and not *in control* (roughly: proximal/distal) that modulates to the different control domains. Other languages are less minimal. Turkish, for

---

[17] The expression must exploit the control profile *subjectively* (see Langacker 2008, chapter 9). To a rough approximation, this means *implicitly*. The control profile can figure explicitly in, for example, a description, like 'Did you see the thing I am holding in my hand', which explicitly exploits a control relation to identify a referent. But this description is not a demonstrative. *Subjective construal* is crucial for a proper treatment of semantics of many sorts of expressions, not just demonstratives. See Langacker 2008 for discussion.

example, has a demonstrative *su*, which is very specifically tied to perceptual control, and can only be used if the speaker, *but not the addressee*, has perceptual control of the referent (Küntay and Özyürek 2006). Ute has demonstrative expressions prescribing specifically that the referent is seen by neither the speaker nor addressee (Diessel 1999, 42). And as noted above in Section 4.4, Ambulas has demonstratives limited to endophoric situations, and hence restricted to attentional (rather than physical or perceptual) control. English has no demonstrative limited to only endophoric anaphora.

4.6 Discussion and comparison with other approaches

Our goal in this section has only been to present an outline of an account of demonstrative semantics that we feel is superior in a number of respects to extant alternatives, and to say enough about it to make it initially plausible. This presentation has not explored all the nuances of the position, nor has it defended it against all potential objections. These latter goals are not feasible given space limitations. Those interested in a fuller discussion should consult Grush and Brovold (in preparation). One point we would like to emphasize is that we are not claiming that exophoric demonstrative reference to physical objects is primary, and other sorts of demonstrative reference (including endophoric) are an extension of this primary sort. While it is true that concept empiricists often appeal to work in cognitive linguistics in their claims that conceptual abilities are based on perceptual abilities, not all cognitive linguists are concept empiricists (for interesting though brief discussion, see Langacker 1993). Our claim is that what is primary is the conceptual archetype of the control cycle. The first applications of this archetype may be in understanding physical control with the ability to apply the archetype to perceptual, attentional, and social domains appearing later. And in fact there is some reason to think that physical control is mastered earlier – Küntay and Özyürek (2006) show that Turkish children don't master the perceptual-control element of *su* until relatively late. But we are not making a concept-empiricist claim, and we are inclined to a position similar to Langacker's. But for purposes of this chapter, we are officially neutral on this issue.

It is worth saying a few words about how this account compares with other approaches.

First, the proximal/distal contrast interpreted spatially is seen to be an imperfect correlate of various aspects of the control profile. This is verified by e.g. the work of Coventry et al. (2008) where manipulations including tool use and who previously touched the referent were studied (see Section 4.2).

Second, Clark et al.'s (1983) work on common ground is accommodated. What Clark et al. term *common ground* is a complex of background assumptions concerning which entities the speaker and addressee are more likely to take attentional or perceptual control of. For example, when a speaker says 'What do you think of that?' when looking at a display case holding both a traditional and a hipster watch, the addressee will interpret the referent differently depending on whether the addressee takes it to be common knowledge that the speaker is shopping for a gift for old conservative Uncle George or young modern Cousin Amanda (see Clark et al. 1983). The work on common ground is not a competing view of demonstrative semantics, but rather a compatible account that fills out details of how the control profile, especially concerning proclivities for attentional and perceptual control, is constructed.

Perhaps the closest extant approach to ours is based on *accessibility* (see e.g. Ariel 1990). To a first approximation accessibility is taken to be 'the ease (of effort) with which particular mental contents come to mind. The accessibility of a thought is determined jointly by the characteristics of the cognitive mechanisms that produce it and by the characteristics of the stimuli and events that evoke it' (Kahneman 2003). There are two ways in which accessibility has been applied to the study of demonstrative semantics. First, it is used to shed light on proximal/distal contrasts. On this front, our view is in agreement with accessibility approaches in moving away from simple spatial interpretations of this contrast. And for the most part, it is possible to see 'accessibility' as a notational variant of certain elements of the control profile, especially proclivities for attentional and perceptual control (though in Grush and Brovold, in preparation, we show some differences). On the other hand, the control profile provides a synthetic model that explains how factors such as purely physical control can influence demonstrative semantics even when more cognitive factors relevant to accessibility, as defined above, are not at issue.

The second area of application of accessibility is in explaining how uses of demonstratives pattern with respect to pronouns and full NPs (see e.g. Ariel 1990; Brown-Schmidt et al. 2005). This is something we have not discussed in this chapter, and we are inclined to take accessibility work on this issue, not as being in competition with our approach, but as compatible work that focuses on issues that have not been the main focus of our account here (choice of demonstratives vs. pronouns or full NPs).

Finally, our approach has overlaps with that of Diessel (1999, 2006), who argues that joint attention is a factor in demonstrative semantics. Though there are differences, notably the fact that the control profile makes appeal to kinds of control other than attentional, and is thus able to synthesize a broader range of data than an approach that focuses on attention.

5   GENERAL DISCUSSION

Our chapter makes three distinct but related points. The first is a point about perception, not language or demonstratives. This point is simply that the idea that *perceptual* processes are keying in on spatial location to isolate its objects appears to be untenable. Appeal to gestalt criteria seems to be a superior approach, both because it covers a wider range of the data and because it subsumes simple location as a special case. The second is a point about the scope of demonstrative reference, not about perception. That point is that perfectly felicitous everyday demonstratives refer to all sorts of things that aren't physical objects, or even spatially located. Putting points 1 and 2 together we conclude that the idea that the mechanisms of physical-object categorization or even location perception are going to play any key role in understanding the semantics of demonstratives is misguided. Combine this with empirical facts to the effect that many languages have demonstratives that *require* the referent not be perceived, and we conclude that the fashionable philosophical notion of the 'perceptual demonstrative' is an aberration.

The third separate point is a positive theory of the semantics of demonstratives. This point is that an account of core demonstrative semantics that is based on something like what we have called the control profile seems well suited to providing a more insightful and general account of demonstrative semantics than the current competitors. The third point is independent of the first and second. If it turned out that perceptual systems used location, or something other than location or gestalt criteria, to isolate its objects, or if it turned out that only entities of some restricted type could be referred to by demonstrative expressions, the points about demonstrative semantics from Section 4 would remain essentially unchanged. It would still be true that one part of the control profile would be perceptual control, and that this would be relevant for some sorts of exophoric demonstratives – there would just be a different psychological story about the details of how the perceptual systems' abilities to gain, maintain, and lose perceptual control.

We close with two observations. First, on our approach, an understanding of demonstrative semantics is a quite interdisciplinary affair. It will involve work that is the home of the linguist, such as knowledge of the spectrum of demonstratives in various languages and the many ways that demonstratives can manifest themselves even in a single language, as well as knowledge of how demonstrative semantics will interact with the semantics of other expressions in discourse and other pragmatic considerations. It will also involve psychological investigations into what the control profile is and how knowledge of it is maintained and shared. Work on joint and shared attention is relevant here, e.g. Tomasello (1995) and Tomasello and Carpenter (2006), as well as studies along the lines of Clark's *common ground*.

Second, for reasons that should be apparent by now, we see no compelling reason to expect much interesting cross-fertilization between, on the one hand, work in psychology and neuroscience on how vision isolates its objects and, on the other hand, the semantics of natural language demonstratives. Some readers of drafts of this chapter have been confused on this point, because in Sections 2 and 3 we argued that gobjects could be accusatives of perceptual processes as well as referents of demonstratives, and this led some to assume that we were claiming that the perceptual isolation of gobjects is connected with demonstrative reference. But we are not.

One way to put this would be by saying that the notion of a 'perceptual demonstrative' being given so much attention in philosophical circles is a misnomer – *even if* we limit (quite unreasonably) investigation to exophoric demonstrative reference in English, perception is only one of many factors. The idea that there is something describable as a 'perceptual demonstrative' that will serve as a fruitful bridge between work on the psychology of perception and the semantics of natural language demonstrative expressions seems to have arisen from a very impoverished view of demonstrative semantics at the linguistic end of things, and a fixation on visual experience of spatial objects on the perceptual end of things. But both of these ends have been false starts.

REFERENCES

Ariel, Mira (1990) *Accessing Noun-Phrase Antecedents*. London: Routledge.
Bahrick, Lorraine E. and Lickliter, Robert (2000) Intersensory redundancy guides attentional selectivity and perceptual learning in infancy. *Developmental Psychology* 36, no. 2: 190–201.
Beun, R.-J. and Cremers, A. (1998) Object reference in a shared domain of conversation. *Pragmatics and Cognition* 6, nos. 1–2: 121–152.

Blaser, E., Pylyshyn, Z. W., and Holcombe, A. O. (2000) Tracking an object through feature space. *Nature* 408: 196–199.

Bregman, Albert (1994) *Auditory Scene Analysis: The Perceptual Organization of Sound.* Cambridge, MA: MIT Press.

Brown-Schmidt, Sarah, Byron, Donna K., and Tanenhaus, Michael K. (2005) Beyond salience: Interpretation of personal and demonstrative pronouns. *Journal of Memory and Language* 53: 292–313.

Calvert, G., Brammer, M., and Iversen, S. (1998) Crossmodal identification. *Trends in Cognitive Science* 2: 247–253.

Campbell, John (2002) *Reference and Consciousness.* Oxford University Press.

Clark, Austen (2000) *A Theory of Sentience.* Oxford University Press.

Clark, H. H., Schreuder, R., and Buttrick, S. (1983) Common ground and the understanding of demonstrative reference. *Journal of Verbal Learning and Verbal Behavior* 22: 245–258.

Coventry, K. R., Valdés, B., Castillo, A., and Guijarro-Fuentes, P. (2008) Language within your reach: Near-far perceptual space and spatial demonstratives. *Cognition* 108: 889–895.

Diessel, H. (1999) *Demonstratives: Form, Function, and Grammaticalization.* Typological Studies in Language 42. Amsterdam and Philadelphia: John Benjamins Publishing Co.

(2006) Demonstratives, joint attention, and the emergence of grammar. *Cognitive Linguistics* 17, no. 4: 463–489.

Ellis, W. D. (1938) *A Sourcebook of Gestalt Psychology.* Kegan Paul, Trench, Trübner.

Gao, T. and Scholl, B. J. (2009) Are objects required for object files? Roles of segmentation and spatiotemporal continuity in computing object persistence. *Visual Cognition* 18, no. 1: 82–109.

Goldberg, A. E. (1995) *Constructions: A Construction Grammar Approach to Argument Structure,* University of Chicago Press.

Grush, R. and Brovold, A. (In preparation) Demonstratives and control.

Henderson, J. M. and Anes, M. D. (1994) Roles of object-file review and type priming in visual identification within and across eye-fixations. *Journal of Experimental Psychology: Human Perception and Performance* 20: 826–839.

Jordan, Kerry E., Clark, K., and Mitroff, S. R. (2010) See an object, hear an object file: Object correspondence transcends sensory modality. *Visual Cognition* 18, no. 4: 492–503.

Kahneman, D. (2003) A perspective on judgement and choice: Mapping bounded rationality. *American Psychologist* 58, no. 9: 697–720.

Kaplan, D. (1989a) Demonstratives. In J. Almog, J. Perry, and H. Wettstein (eds.) *Themes from Kaplan.* Oxford University Press, pp. 481–564.

(1989b) Afterthoughts. In J. Almog, J. Perry, and H. Wettstein (eds.) *Themes from Kaplan.* Oxford University Press, pp. 565–614.

Keetels, M. and Vroomen, J. (2008) Tactile-visual temporal ventriloquism: No effect of spatial disparity. *Perception and Psychophysics* 70: 765–771.

Küntay, A. and Özyürek, A. (2006) Learning to use demonstratives in conversation: What do language specific strategies in Turkish reveal? *Journal of Child Language* 33: 303–320.

Lakoff, G. (1987) *Women, Fire, and Dangerous Things: What Categories Reveal about the Mind.* University of Chicago Press.

Langacker, R. W. (1993) Reference-point constructions. *Cognitive Linguistics* 4, no. 1: 1–38.

———— (2002) The control cycle: Why grammar is a matter of life and death. *Proceedings of the Second Annual Meeting of the Japanese Cognitive Linguistics Association* 2: 193–220.

———— (2008) *Cognitive Grammar: A Basic Introduction.* Oxford University Press.

Lowe, E. J. (2005) How are ordinary objects possible? *Monist* 88, no. 4: 510–533.

Raftopoulos, A. (2009) *Perception and Cognition: How Do Psychology and the Neural Sciences Inform Philosophy?* Cambridge, MA: MIT Press.

Scholl, B. J. (2007) Object persistence in philosophy and psychology. *Mind & Language*, 22: 563–591.

Talmy, Leonard (1988) Force dynamics in language and cognition. *Cognitive Science* 12, no. 1: 49–100.

Tomasello, Michael (1995) Joint attention as social cognition. In C. Moore and P. J. Dunham (eds.) *Joint Attention: Its Origin and Role in Development.* Hillsdale, NJ: Lawrence Erlbaum, pp. 103–130.

Tomasello, Michael and Carpenter, Malinda (2006) Shared intentionality. *Developmental Science* 10, no. 1: 121–125.

Treisman, A. (1996) The binding problem. *Current Opinion in Neurobiology* 6: 171–178.

CHAPTER 3

Visual demonstratives

Mohan Matthen

I INTRODUCTION

Playing a game of squash, my opponent hits a drop shot to the left corner. I run to the front of the court – I do not bump into him or the walls as I do. Then, seeing that the ball has bounced high, I step out on my right foot, and hit the ball high on the front wall for a cross-court lob.

What kind of visual information allows me to plan and then execute this complex act of coordination? I do not mean this as a neuroscientific, or even a psychological, question, but as one in philosophy, and though I hope that what I say will be scientifically sound, I shall be as empirically non-specific as I can.

The question can be posed in an old-fashioned framework. There is a process of practical reasoning that leads up to my action. Minimally:

1. I want to strike the ball.
2. The ball is there.
3. So: I must run there ... etc.

The question is: What premises deriving from visual content must figure in such a process?[1] How do the various terms in the above relate to what we immediately and non-inferentially see?

In this chapter, I distinguish three sorts of ideas that play a role in visually guiding action. My aim is to sketch an account of how these ideas – two types of visual idea and one non-visual (as I shall argue) – interact in visually guided action.

---

[1] See Matthen 1988 and Burge 2005 for treatments (from somewhat different perspectives) of questions like this. Both papers emphasize how the assignment of intentional content to visual states facilitates accounts of visual data processing given by cognitive science.

II   THREE IDEAS OF THE TARGET

For the sake of simplicity, consider just the ball. How is it represented in the above process of reasoning?

1. First of all, I must have a game-related idea of the ball – an idea that gives it a place in the rules and tactics of a squash game – for this governs the formation of intentions such as striking it, doing so before it bounces twice, making it hit the front wall between the tin and the top line, making it difficult for my opponent to strike it, and so on. Speaking more generally, even when one is speaking of the relevance of visual data to behaviour, one needs to bring in non-visual ideas of the objects involved. For when I form the intention to act upon an object, I do so under an idea that fits it into my broader aims. These broader aims are rarely confined to evincing *behaviour* that satisfies certain physical parameters – running in a certain direction at a certain speed, hitting the ball with a certain force at a certain angle, and so on. What I want to achieve is usually comprehensible only in some behaviour-transcending framework. I want to win a point, neutralize the opponent's advantage in court position, trap him against the back wall, and so on. These aims have to be achieved by my bodily movements, but they go beyond these movements. Moreover, my target cannot be specified just in visual terms. I aim to strike that small black sphere, but only because it is the ball in play. It is because the visual specification is identified with the specification of the target under the governing idea that my physical behaviour is launched.

This is true even of animals incapable of explicit reasoning. Consider a dog chasing a ball thrown by its owner. The dog is retrieving something for its owner. If its owner threw a ball, the dog would chase after it; if some other person nearby threw a ball, her dog would not chase it (or can be trained not to). In this case, too, it is because the dog's visual information is assimilated to its broader aims that its physical behaviour is launched. Because the point extends in this way to animal actions, it is not restricted to situations in which highly acculturated terms are in play – as they are in my example of the squash game. Target-oriented behaviour, we might say, presupposes a mental representation of the target under which action on that target is chosen. This mental representation transcends behaviour understood in purely physical terms.

Returning to the squash ball, then, let us label the behaviour-transcending idea of it – the idea relevant to my squash game-related intentions – [B_1]. (When I mean to be talking about the idea, I shall put square brackets around B_1; without these brackets, the symbol denotes the ball.)

2. To translate my game-related intentions into physical behaviour, I must be able visually to identify the ball. Having formed specific intentions with regard to the ball under the idea [B_I], I need to know which object in my vicinity falls under this idea, track where it is, and so on. Thus, I might engage in an implicit mental process something like the following:

I want to strike B_I.
B_I = the thing that looks like so.
So, I want to strike the thing that looks like so.

Let us say that there is a visual idea, or image, that corresponds to the phrase 'the thing that looks like so' above. Call it [B_V]. To know where to direct my intended behaviour, I must visually search for and act on something that satisfies [B_V].

3. Lastly, since physical action is in question, I must possess the information needed to control my body relative to the ball. For this, I must be able to locate and track the ball in 'egocentric space' – by which I mean real space, measured in a coordinate system in which some point on my body is at the origin, $<0, 0, 0>_E$. (The subscript marks egocentric coordinates; action-relevant representation of position is in polar coordinates, so that the first coordinate represents the target's distance, the second its azimuth relative to some body-defined direction, and the third its elevation.) At a certain point I must move my feet in such and such a direction. As I conduct my sequence of actions, I update the position of targets relative to me.

Identifying the ball visually is not sufficient for me to complete my intended course of behaviour. Targeted bodily movement requires me, or my visual system, to compute the position of the ball and other objects relative to my body, i.e. in egocentric space. Once I have done this, I can engage in the following reasoning:

I want to strike B_V.
B_V = the material object at $<r, \theta, \varphi>_E$.
So I want to strike the material object at $<r, \theta, \varphi>_E$.

The upshot of these considerations is that I will have an egocentric-location representation for the ball. Call this [B_E] – let it stand in for 'the material object at $<r, \theta, \varphi>_E$' above. My bodily movement schema at any given point of time will make use of [B_E].

Empiricist orthodoxy maintains that the ball's egocentric coordinates are given in visual consciousness – for, according to a philosopher like

Hume, the 'visual field' is egocentric. This seems to be a mistake. My conscious visual image of the ball and its trajectory is court-centred, not self-centred. I am not consciously aware of its egocentric trajectory – for instance, it does not present the appearance of moving faster in my direction as I accelerate toward it, though in egocentric terms, of course, it is moving faster. I am aware of its speed and my speed relative to the court. I am not conscious of $[B_E]$, it seems – at least not fully.

Hume was wrong, then, about the egocentricity of conscious visual representations; nevertheless, animals perform their tasks in ways that indicate that their movements are controlled egocentrically. As R. S. Woodworth (1899) demonstrated in a remarkable PhD thesis, human voluntary movements are astonishingly accurate. For example, he observed some labourers pounding spikes with a sledgehammer for an hour. He estimated that the arc of their swing was about 150 centimetres (cm); yet their 2-cm-wide target was missed only once in an estimated total of 4,000 swings among them all. For accuracy of this magnitude, he figured, the mean variance of the point of impact would be much less than 1 cm. Similarly, when we rapidly handwrite a row of letters, the corresponding points of the letters – for instance, the top and middle points of the 'b' and of the 'h', and the tails of the 'g' and 'y' – vary in height, Woodworth estimates, by less than 5 per cent. This sort of accuracy in real time demands not only that vision should determine the position of the target, but also that it should provide this information to the efferent systems in the egocentric form that they require. Transforming allocentrically coded conscious vision into egocentric form would be far too slow. $[B_E]$ must, therefore, be computed independently. I'll return to this point in Section IV, below.

III   PROPERLY VISUAL IDEAS

Before I can get to the main problem of describing the interaction among these ideas, I'll need to deal with a preliminary question. Are the three ideas that I have posited really independent of one another, or are they simply three aspects or parts of a unified conscious presentation?[2]

First, I'll defend the view that $[B_I]$ is different from $[B_V]$.

[2] See David Milner and Melvyn Goodale 1995, chapter 1, Andy Clark 2000, Scott Glover 2004, Goodale and Milner 2004a and 2004b, and Mohan Matthen 2005, chapter 13, for background discussion relevant to the next two sections.

Earlier, I argued that [B$_I$] – the idea of the ball that shows the point of the action undertaken – is not, as such, a visual idea. (Here, I am speaking generally: visual recognition tasks and the like might be couched in visual terms.) The ball is acted on under squash-tactical maxims of action. But I need visually to identify the ball to put game-related intentions into effect. And even when [B$_V$] is on-line and governing my actions, the non-visual idea [B$_I$] still figures in my process of reasoning. For suppose that while I am rushing to the ball, I notice that my opponent has retreated to the back of the court. Then, I might change my plan. [B$_I$] is in constant interaction with visual ideas of the ball. In light of these considerations, it seems that [B$_I$] and [B$_V$] are distinct and independent.

Now, the distinction between visual and non-visual ideas has come under attack. The ordinary language verb 'see' does not distinguish between purely visual ideas and ideas that have a non-visual component. I can say I see of something that it is *blue* and with equal linguistic propriety I can say that I see of it that it is a *five-dollar bill*. Yet, the latter requires that I subsume what I see under a non-visual concept, while the former is a visual concept that comes prepackaged with the visual state. This distinction is elided in many philosophical treatments of vision. The notion has become well entrenched in some philosophical circles that there is no structural difference between the two examples given above – that seeing something as *F* is always a matter of having inarticulate visual sensations, which do not in, of, or by themselves present an object as possessing any property. Thus, even seeing that something is blue is taken to require a 'view' or theory about blue things. (For a well-presented recent version of this theory see Anil Gupta 2006.)

Though it is not possible to argue the point in detail here, I regard this notion as utterly mistaken. First of all, there is no such thing as an inarticulate visual sensation – a sensation is always a presentation of some object as possessing some sensory property, and it is so in itself, not in virtue of other beliefs the perceiver holds. (See Matthen 2005, especially chapters 1–3.) We respond instinctively to sensations; contrary to Gupta (2006) we do not have to form a 'view' about what they mean.

Second, it is only in such cases as seeing a five-dollar bill, and not in such cases as seeing something blue, that a further step of subsuming a visual sensation under a learned or inferred non-visual concept is required. For there are some *properly visual* ideas, and seeing a *V*, where *V* is properly visual, is a direct visual apprehension, for which a subject needs no theory. One simple way of making this point is the following, modelled on a move made by Sydney Shoemaker's (1968) and Gareth Evans' (1982) discussion of 'identification-free' judgements (ibid., 179–191). Suppose that you identify a piece of paper poking out of somebody's wallet as a five-dollar bill. There are two ways that your judgement can go wrong. First, you might be wrong about the colour of a Canadian five-dollar bill, and thus you may take a green-looking banknote to be a five-dollar bill. Let's call this an error of *misconception*. Second, the banknote may look blue to you, and since you (correctly) think that all blue Canadian banknotes are five-dollar bills, you may take it to be a five-dollar bill. You are wrong because it is actually green. Let's call this *misperception*. The point that I want to make is that with regard to properly visual concepts, error through misconception is impossible. You can't be led into error by misconceiving blue – by mixing up the visual marks of blue with those of green, for instance. This is the mark of a properly visual idea.

Conclusion 1 – [B$_V$] consists of ideas that are immune to misconception by anybody who can perceive them. [B$_I$] contains ideas that are not immune to misconception. Thus, [B$_I$] ≠ [B$_V$].

IV   EGOCENTRIC VISUAL IDEAS

Now, I'll explore the character of [B$_E$] and show that it is independent of [B$_V$].

Some hold that when you look at the squash ball you get an indivisible package of visual information – that it is round, black, small, moving, and *there*. When you want to act physically on something, the argument continues, you search for it under an incomplete visual description – for instance, you might search for a black, round thing. When you make visual contact with this thing, you (ideally) get a much fuller package of visual information. This package includes its location relative to you. It is only by an act of abstraction that you can separate location out as a distinct visual idea of the ball. [B$_E$] is an inextricable part of [B$_V$].

The intuition that featural and locational information are inextricably linked accords with a notion of eye–hand coordination that some find intuitive. According to this notion, vision provides an image, which the subject must use in order to achieve contact with external things. Think of a video game in which you simulate flying an airplane. You work by monitoring a television image. This image is not used in a purely egocentric fashion – when you shift about in your seat, the image shifts relative to you, but this makes no difference to your piloting. You act on your controls in such a way as to make the image change in certain ways. When an object gets too close, for instance, the imaged gap between the airplane and the object gets too small; you correct this by acting on your joystick in such a way as to make the gap grow bigger. When you want to land the airplane, you manipulate the joystick in such a manner as to make the axis of the runway continue on from the axis of the airplane – the closer end touching the nose – and then you make it get larger and larger while maintaining this alignment. Similarly, it seems, a human subject in the real world can translate his movement plan into an image-manipulation plan. I catch a ball by first moving in such a way as to stabilize the position of the ball image in my visual field, and then I catch it by reducing the gap between the images of the ball and of my hand. The intuition is that bodily movement is controlled indirectly by manipulating $[B_V]$. Let's call this the act-by-image-manipulation (AIM) model. AIM appears to eliminate the need for $[B_E]$.

The AIM model of visual guidance is at odds with developments in cognitive neuroscience. There is now ample evidence that there is no smoothly integrated visual image of the sort envisaged in the intuitive picture of the control of action. In the last few years, the debate around this proposition has revolved around the supposed separation of two visual streams. One of these, the so-called ventral stream, is concerned with characteristics that objects possess independently of the illumination and perspective in which they may stand at the moment of viewing – characteristics such as colour, shape, surface texture, trajectory, allocentric position, etc. The other visual stream, sometimes called the dorsal, is concerned with positional information relevant to the selection and control of behaviour. The neurological component of the distinction is irrelevant to my concerns, and I will not rely on it here.[3] What is just about uncontroversial is that visual processing for perceiver-independent qualities of objects – I'll call this *descriptive* visual processing – is largely independent of processing for egocentric position, which I shall call *motion-guiding* visual processing. My aim is to argue that $[B_V]$, which is furnished by descriptive vision,

---

[3] In Matthen 2005, chapter 13, I made a function-based distinction between descriptive vision and motion-guiding vision parallel to the seminal distinction between vision for perception and vision for action made by Milner and Goodale (1995) and Goodale and Milner (2004a). The functional distinction was meant to be independent of the ventral stream/dorsal stream distinction, also employed by Milner and Goodale. The latter is supposed to be the neurological substrate of the former, but it is conceptually distinct. This is not fully appreciated by some critics. Raftopoulos (2009) criticizes my distinction largely on the grounds that ventral- and dorsal-stream processing are not independent of one another. But I do not intend to make any claim about the anatomical loci of these visual functions. Raftopoulos agrees with my claim that visual reference is determined independently of visual description (ibid., 350), and this is the claim I wish to elaborate here.

does not substitute for $[B_E]$, which is furnished by motion-guiding vision. Later, we'll see how this carves out a special semantic role for $[B_E]$.

Here is a thought experiment that brings out the independence of $[B_V]$ and $[B_E]$. Think of two actions as follows. (A) Suppose that you are sitting in front of your desk, looking at it. There are several objects on it. You pick up the pencil and write something with it. Then, (B) you turn away from the desk, and recall a visual image of it. This time, you select the eraser. You mime picking it up and erasing something.

According to the AIM model, both actions are guided by a visual image similar to one generated by a sophisticated computer game – call this image $[Desk–Hand_V]$ – though in the case of (B), this image might be somewhat degraded. The idea is that in (A), you act on the pencil indirectly by bringing about certain changes in $[Desk–Hand_V]$. There seems to be no reason why you cannot do the same in (B). Of course, action (B) will not be as fluent and accurate as (A). Assuming that you haven't practised (B), in which case you will not be completely reliant on the image, you will not have the same confidence; your reach will be more tentative; it may not be in exactly the right direction or land at the right height; your grip may not be properly sized to the eraser. Why? One reason for this is, of course, that the recalled visual image is not as detailed or accurate as an on-line visual image: this partly accounts for inaccuracies in your action. But there is another important deficiency in simulated action: the feedback that you get from the recalled image does not translate as smoothly into action. This, I believe, is important to understanding the failure to correct for inaccuracies in reaching, and for the slowness and tentativeness of the simulated action.

There are three phases of a *non*-simulated visually guided action such as (A). First, you search for the pencil by its visual characteristics. This first phase is clearly driven by descriptive vision. You use the generic pencil image that you have stored in memory to select an object on your table by its visual characteristics. Next, your knowledge of the object and your intention determines a motor schema – this will include trajectory, speed, style of grip, and force. Finally, vision somehow helps you translate the motor schema into physical behaviour. (As we shall see in a moment, it is not the only guidance unit operating here.)

Now, let's consider the simulated action, (B). In its first phases, this action is very similar. You will select the eraser from the several objects presented by your recalled image, and select a motor scheme that depends on the action you intend to mime. Then, you launch your hand toward the eraser, based on its position in the recalled image – presumably, this image is (or can be) arrayed in front of you in the same manner as a sensory image.

It is in this motion phase that the lack of visual feedback seems most debilitating. But why? You have an image of your desk in your mind's eye. That mental image presents the desk spread out in space. What is missing, relative to (A)? Is it that you cannot see the position of your hand relative to the position of the eraser in your mental image, and so cannot monitor the gap between hand and imagined eraser? But why is this a shortcoming? After all, you possess *bodily* awareness of the position of your hand. And surely this is a large part of the guidance operating in (A).[4] For it is exceedingly unlikely that in the sorts of cases described by Woodworth (1899) – hammering a spike, writing a row of letters – visual feedback is the sole provider of spatial information. Vision seems to provide the position of the spike or the paper, and perhaps some of the last-second course corrections, but it is bodily awareness that tells you how the hammer and the pencil are going, and how much force is needed to complete the action.[5] The spatial awareness provided by vision and by bodily awareness are integrated. So why not AIM in the integrated image?

My intuition is that AIM is actually a good model of how motor schemes are executed when vision is off-line. I would contend that the more fluid execution in case (A) shows that a better method is at work there. On-line vision guides the execution of motor schemes by providing egocentric-location information directly to the limbs. Off-line vision is unable to do this. In simulated action, we are therefore forced to AIM – and since this method is indirect, it is slower and more inaccurate. This leads to a distinction that is (perhaps) a bit crude and overstated. Descriptive visual content contains a message about how things are in themselves, independently of the observer's perspective. This content can be used for movement, but only using AIM. Motion-guiding vision provides the limbs with egocentric coordinates that they can use directly, without the need to translate.

The difference of coding accounts for certain other discrepancies between conscious image and performance in reach-to-grasp manoeuvres.

---

[4] See Santello et al. 2002 and Winges et al. 2003 for evidence in support of this claim. Santello et al. show that simulated actions are deficient in some ways, but can approach visually guided action in other ways.

[5] I practised writing rows of letters with my eyes both open and shut. The shut-eye efforts were nearly as good with respect to uniformity of height and legibility, but quite a bit worse with respect to alignment with the line.

It was shown some years ago that the hand reacts to positional shifts of a target that are not consciously seen (Goodale et al. 1986). When subjects reach for a target that is shifted during a subject eye saccade (just before the reaching action is completed), they can adjust their reach to the new position, even though they report not having seen it. This indicates a non-congruity between the conscious visual image and egocentric positioning. Again, size–contrast illusions – displays in which a target appears larger or smaller than it really is, because of contrasting objects in the display – do not much affect grasping. The target may look larger than it is, but when a subject reaches out to grasp it, she sizes her grip appropriately (Aglioti et al. 1995). Once again, this shows a mismatch between the conscious visual image and the one determining movement toward the target.

It has been proposed that both discrepancies mentioned in the preceding paragraph arise from the differing functions of descriptive and motor-guiding vision. Descriptive vision is concerned with viewer-independent position, and suppresses random and sudden shifts because in all probability they arise from shifts in viewer position. But as far as motion guidance is concerned, it is irrelevant whether these shifts are due to target or viewer movement – either way, they are relevant to the control of movement, and are not to be suppressed as far as this application is concerned. Similarly, descriptive vision is concerned with object size, and uses visual context to discount perspectival variation. By contrast, motion-guiding vision is not concerned with the perspective-independent size of the target, but just with where the fingers are relative to the target. To the extent that the *conscious* visual image derives from descriptive vision, it diverges from what is needed by motor-control functions.

Conclusion 2 – $[B_V]$ and $[B_E]$ are not artificially distinguished parts of a single image. On-line vision uses $[B_E]$ in visual guidance of bodily movement.

V   VISUAL OBJECTS

I argued earlier (p. 47) that vision directly gives us awareness concerning certain objects. I now want to show that some of these are material objects.

This sounds truistic, but it is actually somewhat controversial. For as David Lewis (1966) once wrote (summarizing a 1949–50 report by Roderick Firth): 'Those in the traditions of British empiricism and introspectionist psychology hold that the content of visual experience is a sensuously given mosaic of colour-spots, together with a mass of interpretive judgements injected by the subject' (357). The idea is that vision presents the perceiving subject with such a 'colour-mosaic', which the subject interprets to construct a scene with objects distributed in the external world.

More recently, Austen Clark (2000) has constructed a theory in which the content of visual experience consists of visual features attributed to *places* in a three-dimensional visual field. Clark's visual features are not restricted to colours, and are not spatially minimal. They include 'colour, luminance, relative motion, size, texture, flicker, line orientation' (186). However, the content of visual experience does not include material objects, according to Clark – in his view, visual features are attributed to places, not material objects. 'The characterization of appearance seems to require reference to phenomenal individuals: the regions or volumes at which qualities seem to be located', he says (61) – the regions or volumes are subjects and the features are predicates. Clark uses the term 'feature-placing' to describe the kind of content he ascribes to visual experience.

So, according to both the British empiricists of the Lewis–Firth report and a sophisticated contemporary philosopher of cognitive science steeped in recent neuropsychology literature, material objects are not delivered by vision.

There are good reasons for thinking that the feature-placing view is mistaken. To start with, it doesn't make sense to say that visual features are *predicated* of places. Susanna Siegel (2002) puts the point well in a review of Clark (2000):

> [Clark] repeatedly says that sensory systems attribute qualities to places. For example 'The sensation of a red triangle ... picks out places and attributes features to them' ([Clark 2000] 147; cf. 69, 70, 77, 165, 167, 185). Taken literally, these claims seem questionable. If audition told us that it was a place, rather than something at that place, that was cheeping, we would have all sorts of errors to correct in the move from audition to thought. We would be similarly misled by vision if it told us that a certain region of space was red, while remaining neutral on whether anything occupying that place was red. (137)

Siegel's point is that it is literally false to say that a *place* is coloured, or that it is making a sound (as opposed to saying that there is something coloured in the place or that a sound was emanating from it). It is false, if for no other reason, then because the material object will take its colour and its noises with it when it moves. The colour that resides in a place can change simply because the thing that occupies it is replaced. It is for such reasons that while it may be permissible to say that colours are *in* places, it is not permissible to say that colours are predicated of them.

Figure 3.1 Brightness constancy. Note: The dark stripe on top is equal to the light one in
front. This figure appears with the kind permission of R. Beau Lotto
(see the demonstration at Lotto n.d).

Another point to consider is that visual data processing employs algo-
rithms that work only because they are applied to material objects. That
is, visual data processing would not deliver veridical experience if the
world were not a certain way. Consider the display in Figure 3.1, from
Dale Purves and Beau Lotto (2003, 57).

Light appears to be striking this object from somewhere behind. (Notice
the shadow it casts in front, and the shade that envelops its front side.)
Each stripe looks uniform in colour, but less brightly illuminated on the
front side. Now, it seems clear in the figure that the dark stripes running
along the top of the object are considerably darker than the light stripes
on the front – the former look a darker grey. In fact, as you will discover if
you cover everything else up, the lower parts of the light stripes (the parts
that look as if they are in the shade) are exactly the same brightness as the
upper parts of the dark stripes.

The illusion is explained by noting that the colour of the stripes is computed against the background of assumptions concerning how opaque
objects intercept light. Since the brightness gradient decreases uniformly

in a way that indicates indirect lighting, the visual system delivers experience as of stripes of uniform reflectance. There is no corresponding true assumption about *places* – places are not opaque; they do not intercept light. There is no shade or shadow in a world of places.

Vision is adapted, then, to the contingent presence in the world of things of particular sorts. Zenon Pylyshyn (2007) puts it in this way:

> The mind has been tuned over its evolutionary history so that it carries out certain functions in a modular fashion, without regard for what an organism knows or believes or desires, but because it is in its nature. (ix)

Looking at the Purves–Lotto display (see Figure 3.1), it is clear that, whatever one might *believe* about the natural world, the *visual* world is simply not presented as a world of places. And this is because vision computes brightness as if it is dealing with a world that contains opaque material objects. Elizabeth Spelke (1990) has written that vision identifies material objects by their 'cohesion, boundedness, rigidity, and no action at a distance'. These conditions are characteristic of material objects, and hence they have come to function as principles for the segmentation of scenes into *visual* objects.[6]

Now consider how motion is perceived. Again, Siegel (2002) puts the point well:

> What happens in sensory phenomenology when a subject sees a basketball make its way from the player's hands to the basket? The information that it's one and the same basketball traversing a single path is not given by sentience if sentience is limited to feature-placing. On Clark's view, the information that it's one and the same basketball traversing a single path has to be given non-sensorily. The subject's visual experience stops short. (137)

To emphasize the point, consider the beta phenomenon.[7] A light flashes to your left, goes off, and then another light flashes somewhat over to the right of the first one. If the interval is quite long – five seconds, say – the

---

[6] This leads to certain oddities of visual ontology: vision renders immaterial things such as images in mirrors, shadows, stains, and patches of light as visual objects. These appear as objects because they approximate Spelke's principles most of the time – though since shadows are cast on the nearest object that the light intercepts, they may suddenly expand or contract, and are non-rigid. Cast shadows look object-like, though they are visually distinguishable from material objects, but shade (as in the sides of objects facing away from illumination) does not look object-like.

[7] The phenomenon I am about to describe was called 'beta' by Max Wertheimer, though it is mistakenly called 'phi' in common parlance (and by me in 2005, chapter 12). The rather different phenomenon that Wertheimer called phi is produced by rapid alternating flicker. In phi, we seem to see an occluder moving in front of the flickering lights. This occluder appears to be of the same colour as the background: it is a kind of negative space that appears in front of the light that is flickering 'down' – the momentarily dimmer one. See Psychological Sciences, Purdue (n.d.) for details.

two flashes are seen as unconnected – a flash here, another flash there. However, as the interval decreases, the display is seen as a moving light. In fact, the light is seen as traversing the empty space in between the two flashes. What is the *subjective* difference between the two displays? Clearly, just that in the second case there is an illusory appearance of motion. But places do not move: motion consists of the same thing occupying different places at different times. Thus, Clark, who restricts visual ontology to places, cannot explain how vision can generate the appearance of motion.[8] Yet, as Pylyshyn has long argued, vision not only detects movement but also tracks objects and their features through movement.

This is a world in which most surfaces that we see are surfaces of physical objects, so that most of the texture elements we see move coherently as the object moves; almost all elements nearby on the proximal image are at the same distance from the viewer; and, when objects disappear, they often reappear nearby, and often with a particular pattern of occlusion and disocclusion at the edges of the occluding opaque surfaces, and so on. (2007, x)

What we seem to see in beta is a material object in motion; vision finds it through an application of Spelke's principles.

Finally, there is the evidence of visual perception in infants. Elisabeth Spelke, Renée Baillargeon, and others have observed the orderly emergence of object perception in infants as they grow up. They found results like these:

Infants were found to perceive a partly hidden object as a connected unit if the ends of the object moved together behind the occluder. Any unitary translation of the object in three-dimensional space led infants to perceive a continuous object: Vertical translation and translation in depth had the same effect as lateral translation ... Perception of a moving, center-occluded object was not affected by the object's configurational properties: Infants perceived a connected object just as strongly when the object's visible surfaces were asymmetric and heterogeneous in texture and color as when they formed a simple shape of a uniform texture and color. (Spelke 1990, 33)

If these principles of object segmentation were learned, as was assumed in the empiricist tradition, the pace of their emergence in infants could

---

[8] Clark (2004, 569) responds: 'flow patterns can give a powerful impression of movement (your movement) even though you do not perceive any thing to be moving' – and gives the example of a blur or streak created by a rapidly moving object, in which the object itself is not seen. This response misses the point in two ways. First, whether or not we see an object moving in rapid optic flow, it is undeniable that we *do* see an object moving in the beta phenomenon, and when we look at a ball being thrown. In these cases, we do *not* seem to see a temporal succession of coloured places. Second, it is not necessary to see the thing to which motion is attributed in order to see motion. In the visual blur of a fastball, we see something moving very fast without seeing what it is that is moving very fast.

be expected to be proportionate to the amount of exposure that a given infant has to the relevant data, and the infant's quickness to generalize. And we would also expect that individual humans would arrive at slightly different (though perhaps broadly accurate) segmentation principles – in the way that they arrive at different principles of, say, parallel parking or differentiating between the music of Mozart and Haydn. But they do not. Object segmentation emerges at more or less the same age in all infants, and the principles are the same from one to another. This is evidence that they are innate, and their emergence a matter of ontogeny and development rather than learning.

Conclusion 3 – Vision delivers direct awareness of material objects.

VI VISUAL REFERENCE

Visual states are about *individual* things. And this creates a puzzle. Suppose I am in a darkened room, looking at an illuminated blue sphere – call it S_1. Later, I am taken to another darkened room, and I look at another illuminated blue sphere, S_2. Now it may be that since these two spheres look just the same, I have indistinguishable visual experiences in the two rooms. Yet it seems that in the first room my visual states were targeted on S_1, while in the second they are targeted on S_2. How does it come about that subjectively indistinguishable visual states can be directed toward different objects? Note that an image, such as a photograph, does not change its reference in the same way – it is targeted on the same individual regardless of where I look at it.

One possibility is that visual states single out their objects in a purely descriptive way. Suppose I seem to see a red disc at place p. According to the descriptivist theory, the thing I see is that which most closely resembles what I seem to see. In other words, the object of my visual state is that which most closely satisfies the descriptive content of my visual state. According to this theory, the content of my two visual experiences is 'blue sphere in front of me' or possibly 'blue sphere in front of me that is causing me to have this experience'. In the two rooms, different objects satisfy this description. The two experiences are the 'same' because they both have this content – but the content is satisfied by different referents in different situations.

The descriptivist theory cannot properly accommodate misperception. Let x be the thing I see. I may misperceive x – suppose that x is orange, and that I misperceive it as red. Then, nothing satisfies the descriptive content of my visual state. The descriptive theory would then imply that I perceive nothing (or, if formulated so that I perceive whatever comes closest, that I see a nearby red thing, even though this other thing has nothing to do with my visual state). But this seems wrong: it is x I see, even though I misperceive it.

It is worth noting that all of the visual and non-visual ideas that we have been discussing so far are subject to error in this manner. One can be wrong about an object being a squash ball; one may misperceive its colour and shape; one may be wrong about where it is in terms of its egocentric coordinates.

What makes x the thing that I see, if not that it satisfies the descriptive content of my visual state? In a classic article, H. P. Grice (1961) argued that x is what I see because it causes my visual state (in the right kind of way). This theory offers us an initially satisfactory result: it allows that even things that are radically misperceived could be the objects of our perceptual states. This result runs counter to descriptivist theories in the right kind of way.

However, I am thinking of visual states as reason conferring: states that rationally lead to *thoughts* and beliefs. Let us say that:

> A visual state V is about x if and only if V *directly* and *by itself* gives the perceiver grounds for believing something about x – in brief, if x is a *direct epistemic target* of V.

I will argue that Grice's approach does not always give us the right result concerning direct epistemic targets. Grice may be correct in his analysis of the locution 'S sees x', but if so, it would follow that what one sees is not always the direct epistemic target of one's visual state.

Under what conditions does a visual state constitute grounds for a perceiver to have a thought about an object? In the case of misperception described above, something looks to me as if it is red. That it looks this way to me gives me a defeasible reason for thinking that it is a red disc. The visual state that I have described puts me in direct contact with x for epistemic purposes, though it is not accurate as far as the colour of x is concerned. And it may be that in this particular case, Grice's theory works – the object to which I gain direct epistemic access by means of my visual state is, as it happens, the thing that I see – that is, the object that caused me to have my visual experience.

But now think of a different kind of case. Suppose that I am looking at a red button and its reflection in a mirror. The image is not a physical object, and it has no causal power. It causes nothing. All the information that my eye receives comes from, and is caused by, the real button. According to Grice's theory, then, I see only one thing here, the real button, though I see it twice. However – and this is my point – my view of the image does not directly and by itself give me reason to believe of the real button that it is red. It *directly* gives me reason only to believe that the *image* is red. My view of the image gives me reason for believing that the button is red only in the presence of further beliefs about mirrors and images.

Here is another kind of example. You are in a cloud of fruit flies. You see hundreds of little specks. In Grice's view, you see each and every fly, because each causes some part of your visual image. But this visual state does not give you a reason to believe of any particular fly that it has any particular property. Let's suppose that the flies all look yellow. This gives you reason to believe that every fly is yellow. But in my view, this still does not give you a reason to believe of any particular fly that it is yellow. The reason is that you cannot visually single out any particular fly. You can form beliefs about a particular fly on the basis of how things look to you only if you can visually single it out.

Here is a way of thinking about direct epistemic targets that is consonant with these observations.

> A visual state *V* has *x* as its direct epistemic target, only if *V* directly and by itself enables the perceiver visually to attend to *x*.

One cannot form a perceptual belief about an individual based on a visual state unless one attends to that individual. It follows that taken by itself a visual state can give a perceiver unmediated grounds for believing something about an individual only if it enables the subject to attend to that individual.

This gives the right kind of result in the case of the indistinguishable blue spheres. Each visual state enables me to attend to the sphere that is in fact in front of me, and not the other one. It gives the right result also about misperception: one may well be able to attend to an object despite being mistaken about its colour. Finally, the proposal is designed to deal with the fruit-fly case. You can visually single something out only if you can attend to it. Attention gives the condition under which one can, not only receive information from an object, but also use that information to arrive at beliefs about the individual.[9]

---

[9] This thesis is broadly consonant with John Campbell's (2002) treatment of visual reference.

Now, being able to attend to something is, among other things, a physical capacity. It depends on the subject's ability to turn his eyes to the thing, fixate it, focus on it, etc. By Conclusion 2 (of Section IV), it follows that vision controls attention through egocentric-location coordinates. Of course, it can do this in error; in the case of a stick partly under water, for instance, vision may direct your attention to a location that the stick does not occupy. But this is immaterial: attention is to the thing, not the location. The point that I find important here is that a first condition for forming beliefs about things on the basis of vision is that one is able physically to react to it.

Conclusion 4 – The direct epistemic target of a visual state, X, is that to which the egocentric coordinates $[X_E]$ direct your attention.

VII INDEXICALITY

Paul Snowdon (1981, 1990) makes a suggestion about direct epistemic targets that has a great deal in common with the one advanced in the preceding section. Snowdon proposes that when you visually perceive something, you are thereby capable of making a demonstrative judgement about it. Vision cannot give you reason to believe something about a particular object, unless it bestows upon you, directly and by itself, the ability to single the thing out and attend to it. This is a *physical* ability cognate with the ability to point to the thing, move toward it, and make a demonstrative judgement about it.

I want to flesh out this suggestion in a way that ties it to the visual ideas discussed earlier (Section II). My addendum to Snowdon's suggestion is that vision singles out its object by furnishing the perceiver with an egocentric location for that object. The location is not provided 'explicitly' – that is, seeing something does not enable a perceiver to *say* where things are relative to her. Rather, seeing something enables a perceiver to attend to the thing and orient herself relative to it.

Egocentric coordinates are indexical. They determine a particular position in space, given the perceiver's own position. For any object that the perceiver sees, vision specifies egocentric coordinates for the object. These egocentric coordinates enable the perceiver to direct attention to the object When I look at my computer, I see *it*. At home, I may have an exactly similar visual experience, because I am editing the same document on an exactly similar computer. Yet it is this computer I now see, not the one at home. This is because the egocentric coordinates that my visual system gives me for the computer are indexed to my current location, not my home location. Thus, visual reference is not purely descriptive – it is indexical. (With a photograph, it is different: images presented in a photograph are not indexed to my current location, and looking at the photograph does not enable me to orient myself with respect to the object depicted by it.)

Now, only objects in real space can be assigned egocentric coordinates. Objects such as the 'stars', or phosphenes, that appear when we receive a blow to the head, after-images, etc., have no position in space. Hence, they cannot be demonstrated. An after-image has no position; hence its position cannot be indicated. Even if such things appear, in some sense, to be in front of one, they do not look as if they are in external space. (See Siegel 2006 for discussion relevant to this point.) I'll summarize this position by saying that these private phenomena have only *phenomenal position*, and no egocentric position in the sense intended. I mean thus to acknowledge the appropriateness of positional relations such as 'to the left' etc. for images, but to distinguish these relations from those that imply location in space. After-images are seen as 'to the left' etc., but not as occupying any position relative to me, having any size relative to the size of my body, not as moving relative to my body, etc.

Since an after-image is not an object that occupies space, it makes no sense to ask whether after-images are the same as other objects outside of space. Suppose you are suddenly dazzled by a bright light and so are afflicted by an after-image for a few minutes. After a minute or so, somebody asks you: is the pink spot you now see the same as the pink spot you saw a minute ago? There is no good answer to the question as asked – the after-image has not shown spatio-temporal continuity (since it occupies the same position in your visual field despite your own motion), but on the other hand it has, in some sense, persisted. After-images have no location. So though it makes sense to ask whether the disturbance in your visual field is continuous and located in the same visual-field place, it does not make sense to ask whether it is the same *object*. There *is* no object here, and no appearance of one. As Snowdon (1981) says, we cannot demonstratively identify after-images and the like – 'only objects, so to speak, in the world can be so identified' (190). With such phenomena, things may look to the perceiver as if there is a spot of light or floating spot in front of him, but there is no object about which he can form a belief, and hence no epistemic target of his visual state.

Only on-line seeing *directly* gives you egocentric location in this sense. As argued in Section IV, neither recollection nor mere imaging is capable of guiding bodily motion. This does not mean that only on-line seeing is targeted on objects. If a state is directly descended, or created, from a state that directly gives you egocentric location, then it is targeted on the object given in the ancestor state. For example, when I try to imagine what my daughter would look like in a blue raincoat that I am thinking of getting her as a birthday present, the image I conjure up is targeted on my daughter even though it does not assign egocentric location to her. However, this state does not allow me to attend to my daughter, or gather information about her.

Conclusion 5 –There is an element in on-line seeing (as distinct from recalling, imagining, etc.) that indexically links visual states to external objects; this fails for internal objects such as phosphenes and after-images.

VIII   DISJUNCTIVISM

Snowdon (1981, 1990) endorses a view known as disjunctivism on the basis of his view about demonstratives. I have endorsed the position concerning demonstratives. I will conclude with a critique of disjunctivism (DIS).

Here is the position that Snowdon advances:

> DIS. The best theory for the state of affairs reported by 'I seem to see a flash of light' is that EITHER there is something I can demonstratively identify that looks to me to be a flash of light OR it is to me as if there is something that I can demonstratively identify that looks to me as if it is a flash of light (but there is not).[10]

By saying that this is the 'best theory', Snowdon means to imply that the two disjuncts specifically describe different kinds of situations, each of which would be correctly, but non-specifically, described by 'I seem to see a flash of light.' Thus, neither disjunct can be deleted from the definition without sacrificing completeness.

Snowdon offers us an interesting example (which he takes from J. N. Hinton) to make his point. I am sitting in a darkened room and I seem to see a very brief faint flash of light.

Consider, first, a case in which:

a. There is really a light that I see.

[10] This wording is a composite assembled from Snowdon (1981, 184–185).

In this case, there really is something that looks to me as if it is a flash of light. It has egocentric location, moreover, and I can demonstratively identify it. So, DIS (above) works: the first disjunct is satisfied.

Now consider a different case. Suppose that:

b. My visual presentation is *as of* a light; i.e. it is to me as if there is a light. However, there is no light there – it is an after-image.

Theorists opposed to disjunctivism think that in (b), my visual state is exactly the same as in (a) where what I see really was a flash of light. So, they say, this too is a case in which I seem to see a flash of light. And in this case, too, these non-disjunctivists say, there is something that looks to me like a flash of light – namely, the after-image. Thus, the non-disjunctivists argue, the first disjunct is the best theory of both cases, and there is no reason for the second disjunct to be added in.

Snowdon disagrees with this. One simple way of arguing the point (not Snowdon's, but he would agree with the premises) is this. It is not possible, as I argued in the preceding section, demonstratively to identify an after-image. In case (b), therefore, there is no possibility of demonstrating the object that looks to me as if it is a flash of light. Thus, Snowdon argues, it is not the same experience as in case (a), which *does* support demonstrative identification. This is his reason for thinking that there are two quite different states of affairs in which 'I seem to see a flash of light' is true – one kind supports egocentric location and the other does not. Snowdon says:

The disjunctive picture divides what makes looks ascriptions true into two classes. In cases where there is no sighting they are made true by a state of affairs intrinsically independent of surrounding objects; but in cases of sighting the truth-conferring state of affairs involves the surrounding objects. (1981, 186)

Looks-judgements are made true by two types of occurrence: in hallucinations they are made true by some feature of a (non-object-involving) inner experience, whereas in perceptions they are made true by some feature of a certain relation to an object, a non-inner experience (which does not involve such an inner experience). (1990, 130)

I agree with Snowdon about a number of aspects of this case. But it seems to me that he is wrong about case (b). Translating it into my terms, I would say if the experience in case (b) is genuinely *as of* a light – if it genuinely is to me as if there is a flash of light – then it should seem to me as if the light is in a certain position relative to me. In my way of thinking about this matter, it is characteristic of seeming to see an object that

that object is endowed with egocentric, not merely phenomenal, location. Consequently, my visual state in (b) will *seem* to support pointing, moving towards, etc. – though *in fact* it does not support pointing, moving towards, etc. Thus, my visual state *does* support a demonstrative – that is, it assigns egocentric location to the flash of light – but because my visual state is inaccurate, the demonstrative that it supports is vacuous, and does not single anything out.

It is instructive, here, to consider two further cases:

c. My visual presentation is *as of* an after-image; it is to me as if I am suffering an after-image. However, I am not suffering an after-image; it is a real light that I see.

Here, it seems to me, my visual state fails to assign egocentric-location coordinates to the thing that I see. If it seems to me as if it is an after-image, then it seems to me as if it isn't a real thing in the external world, and hence it is not sensed as having real location relative to me, only phenomenal location. In this sense, it is experientially like:

d. My visual presentation is as of an after-image; it is to me as if I am suffering an after-image. And I am indeed suffering an after-image.

Snowdon maintains (1981, 190) that 'a person seeing a light but believing that he is having an after-image *may be allowed to make a demonstrative judgement to the effect that that is an after-image*' (my emphasis). I take it that he means that case (c) supports a demonstrative. Thus, Snowdon thinks that in case (c) I will have judged of the light that it is an after-image. This is where I disagree: phenomenal location does not even seem to support demonstrative identification. Therefore, I am unable to make a judgement about the light.

Here, then, is the difference between Snowdon's position and mine. Snowdon thinks that demonstrative thought is impossible without something that is demonstrated. Thus, he thinks that the condition under which a visual state is demonstrative is an external condition – i.e. whether there is a light there. I think that a visual state is demonstrative if it assigns egocentric coordinates to a thing. This is an internal condition. On my way of thinking, demonstrative thought is vacuous when something seems to possess egocentric location, but does not. Further, Snowdon thinks that after-images appear to have egocentric location, and that demonstration is not only possible but successful when there is something at the egocentric location that the after-image appears to have. I believe, on the contrary, that part of what it is to seem to see an after-image is to seem to see something that has no egocentric location.

In my view, it is (a) and (b) that present things as possessing egocentric location, while (c) and (d) do not. This view is different from the one that Snowdon is most concerned to oppose. His main target is the view that there is some core experience common to all four cases. I join him in rejecting that view. On the other hand, I reject the position that, as Snowdon puts it, 'it is quite possible for elements (objects, or states of affairs) external to the subject to be ingredients of an experience' (1990, 124). This leads him to the view that (a) and (c) are genuine demonstratives, and (b) and (d) not. My view, to repeat it once again, is that demonstration requires only something that looks like an object and which vision endows with egocentric coordinates.

IX   CONCLUSION

I have argued that on-line visual states assign seen objects egocentric locations. It is by means of these location assignments that perceivers act on these objects quickly and accurately. Egocentric-location assignments also enable the perceiver to attend to the objects, and thus to form beliefs about them on the basis of nothing other than the visual state itself. Off-line visual states such as recalling and imaging do not assign seen objects egocentric locations, and do not directly support physical movement or information gathering about any object. Moreover, subjective visual phenomena such as after-images are also not assigned egocentric coordinates. This is why these phenomena do not seem to be presentations of real external objects.

REFERENCES

Aglioti, Salvatore, DeSousa, Joseph F. X., and Goodale, Melvyn A. (1995) Size contrasts deceive the eye but not the hand. *Current Biology* 5: 679–685.

Burge, Tyler (2005) Disjunctivism and perceptual psychology. *Philosophical Topics* 33: 1–78.

Campbell, John (2002) *Reference and Consciousness*. Oxford: Clarendon Press.

Clark, Andy (2000) Visual experience and motor action: Are the bonds too tight? *Philosophical Review* 110: 495–519.

Clark, Austen (2000) *A Theory of Sentience*. Oxford: Clarendon Press.

(2004) Sensing, objects, and awareness: Reply to commentators. *Philosophical Psychology* 17: 563–589.

Evans, Gareth (1982) *The Varieties of Reference*, edited by John McDowell. Oxford: Clarendon Press.

Firth, Roderick (1949–50) Sense-data and the percept theory, pts. 1 and 2. *Mind* 58: 434–465, and 59: 35–56.

Glover, Scott (2004) Separate visual representations in the planning and control of action. *Behavioral and Brain Sciences* 27: 3–24.

Goodale, M. A. and Milner, A. David (2004a) *Sight Unseen: An Exploration Of Conscious and Unconscious Vision*. Oxford University Press.

(2004b) Plans for action. *Behavioral and Brain Sciences* 27: 37–39.

Goodale, M. A., Péllison, D., and Prablanc, C. (1986) Large adjustments in visually guided reaching do not depend on vision of the hand or perception of target displacement. *Nature* 349: 154–156.

Grice, H. P. (1961) The causal theory of perception. *Proceedings of the Aristotelian Society*, Supplementary Volume, 35: 121–152.

Gupta, Anil (2006) Experience and knowledge. In Tamar Szabó Gendler and John Hawthorne (eds.) *Perceptual Experience*. Oxford: Clarendon Press, pp. 181–204.

Lewis, David K. (1966) Percepts and color mosaics in visual experience. *Philosophical Review* 75: 357–368.

Lotto, Beau R. (n.d.) Illusion demos. In *Lottolab Studio*. Website. www.lottolab.org/illusiondemos/Demo%2016.html

Matthen, Mohan (1988) Biological functions and perceptual content. *Journal of Philosophy* 85: 5–27.

(2005) *Seeing, Doing, and Knowing: A Philosophical Theory of Sense Perception*. Oxford: Clarendon Press.

Milner, A. David and Goodale, Melvyn A. (1995) *The Visual Brain in Action*. New York: Oxford University Press.

Psychological Sciences, Purdue (n.d.) Phi is not beta. In *Psychological Sciences*. Website. Purdue University. www1.psych.purdue.edu/Magniphi/PhiIsNotBeta/phi1.html

Purves, Dale and Lotto, Beau R. (2003) *Why We See What We Do: An Empirical Theory of Vision*. Sunderland, MA: Sinnauer.

Pylyshyn, Zenon W. (2007) *Things and Places: How the Mind Connects with the World*. Cambridge, MA: MIT Press.

Raftopoulos, Athanassios (2009) Reference, perception, and attention. *Philosophical Studies* 144: 339–360.

Santello, Marco, Flanders, Martha, and Soechting, John F. (2002) Patterns of hand motion during grasping and the influence of sensory guidance. *Journal of Neuroscience* 22: 1426–1435.

Shoemaker, Sydney (1968) Self-reference and self-awareness. *Journal of Philosophy* 95: 555–567.

Siegel, Susanna (2002) Review of *A Theory of Sentience*, by Austen Clark. *Philosophical Review* 111: 135–138.

(2006) Subject and object in the contents of visual experience. *Philosophical Review* 115: 355–388.

Snowdon, Paul (1981) Perception, vision and causation. *Proceedings of the Aristotelian Society*, n.s., 81: 175–192.

   (1990) The objects of perceptual experience. *Proceedings of the Aristotelian Society*, Supplementary Volume, 64: 121–166.

Spelke, Elizabeth S. (1990) Principles of object perception. *Cognition* 14: 29–56.

Winges, Sara A., Weber, Douglas J., and Santello, Marco (2003) The role of vision on hand preshaping during reach to grasp. *Experimental Brain Research* 152: 489–498.

Woodworth, R. S. (1899) *The Accuracy of Voluntary Movements*. Psychological Review Monographs. London: Macmillan.

CHAPTER 4

Losing grip on the world: from illusion to sense-data

Derek H. Brown

During an illusory perception in which I mistakenly perceive a blue car to be purple, what I am wrong about – the mistaken aspect of the perception – is the car's colour. My core interest is in whether or not a particular kind of successful reference, acquaintance, should be applied to this error. One analysis, famously endorsed by indirect realists, holds that in making this perceptual error one is acquainted with something other than or in addition to the blue of the car, say a purple sense-datum (i.e. a subjective perceptual object). Another analysis, familiar from the intentionalist variety of direct realism, asserts that, if anything, one is acquainted with the blue of the car, but one is experiencing it to be purple because one is merely representing purple, or is in an unsuccessful purple-representing state. On this view in no sense is one acquainted with some purple object that exists over and above the blue car. In illusory experiences, when considering the misperceived aspect of the experience, is there successful or unsuccessful reference? This is what is at issue.

The claim that perceptual illusions can motivate the existence of sense-data is both familiar and controversial. My aim is to carve out a subclass of illusions that are up to the task, and a subclass that are not. It follows that when we engage the former we are not simply incorrectly perceiving the world outside ourselves, we are directly perceiving a subjective entity:

Early stages of this chapter benefitted from funding from the Vice-President Academic at Brandon University and the assistance of Wes McPherson. Earlier versions prompted helpful reactions from audiences at my keynote address at the University of Regina and University of Manitoba Graduate-Student Symposium (27 March 2010), and at the Bucharest Colloquium in Analytic Philosophy (3–5 June 2010). The paper also owes a debt to Thanos Raftopoulos' comments. A considerably shortened version was more recently presented at the American Philosophical Association meetings (Pacific Division) in San Diego (20–23 April 2011). Unfortunately, due to publishing constraints I am unable to include discussion of the issues that arose from that session, but wish to thank the audience for a stimulating discussion and especially my commentator, Maja Spener, for many helpful insights.

one's grip on the external world has been marginalized – not fully lost, but once-removed. However, admitting that various illusions do not give evidence for sense-data considerably limits the power of the argument from illusion (Section 6) and brings out its distinctness from the argument from perceptual relativity (Section 7). To reach these conclusions we will examine the role of ambiguity in perception (Section 3), its connection to illusion (Section 4), and the link reference has to every element of this discourse (Sections 2 and 5).

Reference takes center stage because at the heart of our discussion is a form of acquaintance which consists of a variety of perceptual, non-linguistic, reference. Linguistic reference is in my view mostly a distraction here, in large part because the publicity and multivocality of public terms relevant to perception, like 'red' and 'square', have the potential to unnecessarily wreak havoc at all stages of the discussion. I will do my best to steer the reader clear of these traps, and instead trace a clear(er) path to why some illusions support the existence of sense-data.

The inference from illusions to sense-data has been used to additionally argue for indirect realism, the claim that the immediate objects of perception are always (or at least typically) sense-data. On this view our perceptual access to the external or mind-independent world is thus robustly indirect, relegated perhaps to the knowledge acquired through the representational capacities of sense-data. An evaluation of this additional claim is tangential to the present work, although its relevance will be made explicit (Sections 1, 6, and 7), if only to help frame the significance of our discussion.

I THE DEBATE

Ours is a debate between indirect and direct realist approaches to perception, where the latter asserts that we routinely directly perceive the world, and the former that we perceive the world through the aid of a subjective mediating object of perception, something I will call a *sense-datum*. In the terminology I employ, indirect realists believe that when we perceive the world outside ourselves, we do so by first being *perceptually aware* of subjective sense-data, and, in virtue of those data accurately *representing* the objective world, thereby perceive that world.[1]

---

[1] I take 'perceptual awareness' to inherently involve the first-person state of the perceiver and to be in this sense epistemically internalist: it is something to which the perceiver must have fairly immediate access. 'Perception' generally can be understood to cover both internalist and externalist conceptions of perception, unless otherwise stated. Perceptual awareness is also to be distinguished from

Direct realists hold the much simpler view that perceiving the world outside ourselves only involves being perceptually aware of that world. I hope that the reader does not take this axiomatic simplicity to carry more weight than it deserves. Simpler axioms are on their own preferable but not decisively so, whereas failing to accurately recover the data the axioms are formulated to illuminate is a decisive shortfall. The latter is where our debate has raged for centuries, and where our discussion is to be found.

The only variety of indirect realism that I believe has any merit is *projectivism*. As I articulate the view it asserts that sense-data are reflexively projected by one's mind into what one (preanalytically) experiences as the world outside oneself.[2] Sense-data are not experienced as subjective entities but are instead experienced as items that are in the mind-independent world. An indirect realism that does not posit projectivism is hopeless, for when we perceive we directly perceive the world (we experience as) outside ourselves: when I look at the table I see in front of me I do not experience a subjective intermediary, something in between the table and me; I only experience the table. This is another way of making the familiar point that indirect realism generally is at odds with a well-informed element of common sense that direct realists often use to anchor their view.[3] A projectivist indirect realism preserves this element and hence will be presupposed in what follows.

In focusing on the debate between indirect and direct realism I do not mean that other approaches such as idealism are not relevant to our discussion. I mean only that in my estimation the main thread of what follows is best appreciated by a continued focus on the indirect-/direct-realism debate, a claim I will justify in various places. The particular variety of direct realism that will be of interest is intentionalism;[4] due to length

'cognitive awareness', which I take to include not only perceptual awareness but also non-perceptual thought, such as contemplation. Although the analogy is in some ways misleading, the reader might find it useful to think of the distinction between directly and indirectly perceiving something through familiar examples like live television, where one directly perceives the screen and its features, and in virtue of their accurately representing what they depict, one indirectly perceives those depicted items.

[2] On this view this preanalytic disposition to see sense-data as outside oneself can be overcome but is typically not.

[3] Among other things I am referring to Harman's ([1990] 1997) influential transparency argument. See my article, 'The Transparent Projectivist' (forthcoming), for an explication of how projectivism undermines that argument far better than qualia realism does.

[4] Four of many defenses of intentionalism are Harman [1990] 1997, Dretske 1995, Tye 2000, and Byrne 2001, although Byrne interestingly maintains, in contrast to standard approaches, that his variety is consistent with indirect realism. One can argue (as Hilbert 2004 does) that the contemporary root of this movement is Armstrong 1961.

constraints, other varieties, such as qualia realism and disjunctivism, will only be mentioned in passing.

2 ACQUAINTANCE AT A GLANCE

I am concerned with the extent to which a successful form of perceptual reference, what I call *acquaintance*, is involved in perceptual awareness. My interest is in a *minimal acquaintance doctrine*, according to which a perceiver being acquainted with an object (/property/fact) of perception consists of: a perceiver and particular object that exist (e.g. do not merely subsist) and the holding between them of the fundamental relation of acquaintance or perceptual contact. The acquaintance relation itself consists of a form of basic perceptual knowledge of that object possessed by that perceiver. It has been common to add to this for example the claim that the perceiver has infallible or complete knowledge of an item of acquaintance. I will make no use of such additions and hope the reader agrees that they are inessential to the acquaintance idea.[5,6] Instead, acquaintance is meant to provide for a perceiver an epistemic ground or anchor to an object, one that she can exploit to acquire knowledge of various truths (i.e. knowledge by description) about that object.

All are familiar with attempts to avoid understanding perceptual awareness in this way, for example by adverbialists and more recently by intentionalists – I presuppose the latter. The rough idea is that perceptual awareness need not involve the perceiver being in perceptual contact with an existing particular object of perception, but instead need only

---

[5] On at least one reading, Russell himself (e.g. 1912 [1959], 1913) did not take acquaintance to imply infallibility in any robust sense. When acquainted with an item one knew that item with certainty (knowledge by acquaintance of it), but did not because of this possess knowledge of any truths about the item (knowledge by description of it). Indeed given Russell's own scepticism about the reliability of memory, it is reasonable to hold that one's knowledge by acquaintance of some item need not extend to future perceptions of it. For similar reasons I reject the claims that acquaintance with something entails complete knowledge of it and that acquaintance can only be had with simple (as opposed to complex) objects. Regrettably, my positive characterization of knowledge by acquaintance will be minimal.

[6] I take it as straightforward that this acquaintance doctrine satisfies a robust form of perceptual presence, and assume that 'relationalists' (as discussed e.g. in Crane, 2006) are for our purposes individuals who see something like this minimal acquaintance doctrine as definitive of perceptual states. Relationalists of course come in both direct and indirect realist strains – I will be concerned solely with the latter. Acquaintance is also often taken to be a form of non-conceptual perceptual awareness. With a few exceptions to follow, I wish to remain mute on this issue. I see it working in the background in various places, but would have to greatly lengthen the work to bring them all out. (Note that relationalists like McDowell and Brewer take the relations to the world afforded by perception to be thoroughly conceptual.)

involve the perceiver being in the kind of state that represents such an object. The approach is well established in domains like (non-perceptual) thought, where one can without doubt think about things that do not exist (e.g. vampires). A reasonable explanation of this capacity holds that this is achieved by representing those non-existing things, thus bolstering the idea of representational states that do not refer. It also applies straight-forwardly to other propositional attitudes such as desires, fears, etc. The relevant intentionalist thought is that we should extend this approach to perceptual awareness, so that to be perceptually aware is to be represent-ing something (in the perceptual way). This opens the door for perceptual representational states whose objects do not exist (hallucinations), and for states in which only some of the represented elements exist or are instanti-ated (illusions). My interest is in the latter.

Thus whereas for the acquaintance theorist, being perceptually aware of redness requires that the perceiver be acquainted with an instance of red, for the intentionalist, being perceptually aware of redness requires being in a state that is about or directed toward redness, with no com-mitment to an existing or instantiated redness being perceptually pre-sent to her. This disagreement, as it manifests itself in illusion, is our topic.[7] I will call a perceptual state that is about (e.g.) redness but does not involve acquaintance with an existing instance of redness a *merely representational* state. It is somewhat like the difference between shak-ing hands (acquaintance) and pointing fingers (non-acquaintance): the former cannot be achieved unless the item of interest is present or actu-ally exists, whereas the latter has no such requirement – one can point in the direction in which one takes the item of interest to be, without that item actually being there.

It is common for theorists to be dogmatic on this issue and presume that acquaintance (or non-acquaintance) is central to all perceptual states.[8] Such an acquaintance theorist holds that being in a perceptual state consists of being acquainted with something, and the correspond-ing non-acquaintance theorist holds that being in a perceptual state

---

[7] In case there is doubt, note that sense-data are for the indirect realist *the* primary objects of per-ception, and thus are elements of what is commonly called the 'intentional content' of perceptions (see e.g. Brown 2010). The disagreement we are focused on is therefore not about whether or not there are non-intentional aspects of experience – that is the core dispute between intentionalists and qualia realists (e.g. Block) – it is about whether or not certain elements of the intentional content of perceptual states exist as particulars at the time of perception.

[8] In the case of the disjunctivist, a direct realist, acquaintance is definitive of all veridical perceptual states. By contrast for the sense-datum theorist it is definitive of all perceptual states. Disjunctivists, however, do not agree on how to positively characterize misperceptions like illusions.

consists of being in a representational state of the perceptual sort. These platitudes are a mistake. Methodologically, we should be open to the possibility that some phenomena suggest an acquaintance approach (to them) and others do not. More pointedly, the idea of acquaintance (or non-acquaintance) should not be an operational constraint or basic assumption a theorist subsequently fits phenomena into. The idea should instead be applied when appropriate and withheld when not, and we should be doing our best to mine, from phenomena like illusions, information that can help us make these decisions – or so I will purport to demonstrate.[9]

3 AMBIGUITY

In a perceptual circumstance I will take the *stimulus* or *given* to mean the object or scene as it is currently presenting itself to the perceiving agent.[10] Consider a scenario in which a perceiver finds herself in front of a wire cube oriented with the front face slightly pitched up and to the right (UR oriented), as the Necker cube is sometimes drawn. The cube and this particular way it is presenting itself to our agent is the stimulus. In this case the stimulus is objectively ambiguous – *stimulus* ambiguity – in the sense that various objectively different objects could present themselves to the agent in ways perceptually indistinguishable from the way this cube is now presenting itself to her.[11] An obvious alternative would be a wire cube with the front face pitched down and to the left (DL oriented). Other alternatives include a roughly two-dimensional wire figure (2D figure) whose shape traces a flat drawing of the Necker cube; a stretched cube, that is, a figure with square front and back but horizontally elongated (roughly) rectangular sides, and either UR or DL oriented; and so on. Each such alternative marks a candidate disambiguation of our stimulus.

---

[9] This methodological openness is thus consistent with but does not entail a 'hybrid' approach to perception according to which some elements of perceptual content are subjective and others are not (e.g. Maund 2003; Hellie 2006).

[10] I thus do not mean by 'stimulus' or 'given' the pattern of light reaching the eye, although the term is sometimes used in this way. A stimulus in my sense is objective, consisting of the objects and properties (perhaps also facts) being perceived, along with the ways those entities are presenting themselves to the agent at the time of her perception. Compare with Schellenberg's (2008) situation-dependent (SD) properties and the objects possessing them. Her view will be discussed in Section 7.

[11] Gupta (2006) argues for a robust form of ambiguity in the perceptual given, though he construes the phenomenon as one of functionality instead of ambiguity (and for good reasons). I regret that I am unable to devote any space to his view in the present work but wish to acknowledge his influence on my thoughts.

Perceptual ambiguity requires that the agent *see* or *perceive* a stimulus as ambiguous. In keeping with our example, for our purposes this means that she sees the stimulus one moment as being one thing, say a UR-oriented cube, and at the next as being some other thing, say a DL-oriented cube. She can of course also see the stimulus as being a 2D figure, as a UR- or DL-oriented stretched cube, etc. However, some of these disambiguations are perceived more readily than others. I would venture to say that seeing this stimulus as a UR- or DL-oriented cube is easiest, and that it is roughly equally easy to see the stimulus as being either of these ways. By contrast seeing it as a 2D figure is somewhat more difficult, and as a stretched cube (of either orientation) more difficult still. There may furthermore be other disambiguations of the stimulus that the agent cannot see it as (think for example of Moretti's blocks).

We thus have two distinct dimensions to an account of perceptual ambiguity, one consisting of the candidate disambiguations of the objective stimulus (the *disambiguation dimension*), and the other of the extent to which the agent can see the stimulus in accordance with each of these disambiguations (the *seeing-as dimension*). I suspect but will not argue in detail for the claim that the seeing-as dimension can involve some level of cognitive penetration: the age-relative reactions to the dolphin illusion give decent evidence for this, as does the general fact that with practice/education it can become easier to see an ambiguous stimulus in accordance with various disambiguations. Nonetheless much work is done subcognitively, independently of higher-level cognitive penetrations. The fact that seeing the stimulus in our example as a UR- or DL-oriented cube are easiest, and roughly equally easy, suggests that our subcognitive systems have honed in on these disambiguations and judged them to be the most probably correct ones (and equally probably correct ones). I will generally say that the set of disambiguations an agent most easily sees an ambiguous stimulus as is the set of disambiguations her perceptual system judges (be it subcognitively or through both cognitive and subcognitive mechanisms) to be the most probable disambiguations of the stimulus. Disambiguations that it is harder to see the stimulus as are thus judged to be less probable, and so on.[12]

---

[12] These judgements are likely informed by evolutionary pressures, life learning, and perhaps other elements of the perceptual scene. They are thus to some degree contingent. Stimulus ambiguity is the norm in most perceptual research. One common response is to isolate operational constraints that are used or could be used by our vision system to cut down on the possible disambiguations. With respect to shape perception, familiar constraints include 'objects are rigid', 'objects persist', etc. (see e.g. Spelke 1990); in colour perception constraints might include assumptions about the

4 PERCEPTUAL ILLUSIONS AND AMBIGUITY

I do not have a fully precise or adequate conception of (perceptual) illusion on offer. I instead put forth a conception determinate and adequate enough for our purposes. Illusions and hallucinations both involve some form of perceptual error. But they are different. When one hallucinates, one experiences something (or has an experience as of something[13]) that does not exist: if one hallucinates one's mother's voice one seems to hear one's mother's voice when in fact no voice is present. By contrast an illusory experience involves in some way *misperceiving* something that does exist, as when in twisted-cord illusions one experiences, as bent, cords that are in fact straight.

We are interested in a wider and a narrower conception of illusions. The wide one holds that all misperceptions are illusory. The narrower one holds that some misperceptions are mere errors as opposed to illusory experiences, perhaps reserving the term 'illusory experiences' for misperceptions involving a robust form of deception or sensory trickery.[14] Adherents of the narrower conception demand that for a stimulus to be considered an illusion it must be capable of producing misperceptions more than by accident, and ideally in most or all (human) perceivers. Thus the Müller–Lyer illusion continues to generate misperceptions even once we know and perhaps to some extent can see that the two lines are of equal length. Further, we can imagine the narrow advocate wanting to exclude some

composition of common light, and so on (see e.g. Wandell [1989] 1997). These constraints are presumed (by this author and others) to operate subpersonally in an intermediate stage of visual processing and to be at least largely impenetrable by higher-level cognition (see e.g. Raftopoulos 2009, 2011). The vision system applies them to illumination information retrieved by the retina (in early vision) and computes or 'judges' which disambiguation(s) represent the most probable objects of (i.e. external objects causing) that perceptual state. What I suggest follows squarely in this framework, although I am not committed to the variety of cognitive impenetrability that e.g. Raftopoulos defends. Raftopoulos sympathizes with the idea that which disambiguation a perceiver sees a stimulus as is not under direct control of higher-level cognition. Instead 'there are crucial points [on the stimulus] fixation on which *determines* the perceptual interpretation' (2011, 12; emphasis added). On this approach, higher-level cognition may exert some control over where the perceiver's gaze is focused, but not on what disambiguation this triggers the perceiver to see the stimulus in accordance with. Thus 'the cognitive [or higher-level] effects influence the way the figure is perceived only in an indirect way' (ibid.). As is well known, there is *some* evidence for this kind of 'triggering' (e.g. Hochberg and Peterson 1987) but I, for one, doubt that *all* instances of seeing an ambiguous stimulus in accordance with one disambiguation and then another can be so explained. Regrettably I cannot dwell on the matter here.

[13] These phrases are one way to differentiate between acquaintance and non-acquaintance approaches to illusions and hallucinations. For now I am non-committal on the matter, but that will soon change.

[14] Crane adheres to the wide conception, for according to him 'illusion … need not involve deception' (2006, 132). Smith is also a plausible candidate (2002, esp. 23).

misperceptions from the class of illusory experiences, perhaps an isolated case in which I misperceive my sister's voice to be my mother's. The narrower conception is limited because an explication of sensory trickery is difficult to muster, because the line between stimuli that cause misperceptions by accident versus by trickery is vague, and so on. Although I prefer the narrower one, my aim is not to defend it, but instead to explain the impact these different conceptions have had and should in future have on some key philosophical disputes.

A concrete connection to illusions is now possible, for many illusions are intimately connected to perceptual and stimulus ambiguity. On the wide conception of illusions, all misperceptions due to stimulus ambiguities are illusory. On the narrow one, stimulus ambiguity that gives rise to misperception does not on its own entail sensory trickery or therefore illusion. Misperceptions like mistakenly seeing the wire cube as DL oriented are arguably not due to any kind of trickery, but simply to mistaken judgement about a paucity of object information, and hence can reasonably be deemed non-illusory in the narrow sense. However, narrowly illusory perceptions due to ambiguity do exist, for an ambiguous stimulus can contain enough cues to prompt one's vision system to judge that an incorrect disambiguation is most probable, and as a result, trick the perceiver into reflexively seeing the perceptual object as being that incorrect way. An example is 'the Mysterious Floating Vase'. One reflexively sees this as a floating vase, *despite* the otherwise improbability of such an event, when in fact the vase is sitting on the perceived surface and an elliptical dark patch has been added to trick the vision system into interpreting it as a shadow projected by the vase and hence the vase as floating. The shadow cues are so strong that it is difficult to view the scene in the veridical way, but one can do it with some effort, and when one does so the ambiguity of the stimulus becomes apparent (more on this below). That there is sensory trickery here is beyond dispute, and because of that this ambiguous perception should be regarded as illusory even in the narrow sense.[15]

---

[15] For simplicity I will assume the dark patch is paint as opposed (say) to a shadow cast by something other than the vase. 'Floating illusions' can be created in other ways. E.g. one can suspend an object by invisible wires or occluded rods, place it on transparent glass, and so on. There are many examples available in the public domain. My source for the floating-vase illusion is Seckel (2003), which I have regrettably been unable to get permission to reprint. The floating man, to which all that follows is directly applicable, is readily available online. Note also that many other illusions work because of stimulus ambiguity. The old–young woman is a classic, and the dolphin illusion, one of my favourites. The difference is that for these last two examples the ambiguity concerns what these images or drawings represent, not (as in the vase and man case) the fact that distinct objects can present themselves in indistinguishable ways. The latter is our focus.

Another ambiguous stimulus is the famed tilted penny, where the stimulus is at least geometrically ambiguous between an elliptical object being viewed head-on (elliptical disambiguation), or a round one being viewed at an angle (round disambiguation). In this case (like the floating vase and unlike the wire cube) the vision system does not treat both disambiguations as equiprobable but instead favours the round one; we tend to see this stimulus as a tilted penny instead of as an untilted elliptical object.[16] There are many reasons we can offer for why this preference obtains (e.g. the object has other penny features such as being copper in colour, having a head stamped on its side and so on, and pennies are round; our environment has far more round things than elliptical things; etc.). The important point for our purposes is that the preference does obtain, and that *it obtains independently of the actual stimulus in a given case*. If, as supposed, the object *is* a tilted penny then one's seeing it as such is accurate, no sensory trickery is present and hence no illusion (in the wide or narrow sense) should be ascribed. However, if in another circumstance the object is in fact elliptical and untilted – an appropriately oriented *faux penny* – then one's vision system would still prefer the round disambiguation and hence one would be prompted to incorrectly see it as a tilted penny. This misperception is arguably arising because of misleading cues, for it takes a rather special (given our environment) object being oriented in a rather specific way to prompt the misperception. One could thus argue that it is illusory in both the narrow and wide sense.

In this respect I partly agree and partly disagree with Smith's and Schellenberg's analyses. Both correctly assert that a perception of a tilted penny does not constitute an illusory experience.[17] In the past various thinkers have held otherwise, and their mistake was a failure to recognize what has just been stated: (a) that stimulus ambiguity need not yield perceptual illusion, for various disambiguations of stimuli are given a zero or low probability rating by our vision systems and hence not even perceivable by us; and more directly (b) that perceptual ambiguity only yields illusion in the wide or narrow sense when misperception occurs, and we do not usually misperceive a tilted penny to be an untilted elliptical object.

---

[16] This phenomenon is connected to shape constancy, a connection that has been removed to meet length restrictions (but see Brown, under review).

[17] '[I]n no sense, not even in the extended sense given to the term in these pages, is the look of such a tilted penny an illusion' (Smith 2002, 172). See Schellenberg 2008, esp. 74–75, and Section 7 of this chapter.

However, Smith's and Schellenberg's claim that viewing a tilted penny does not constitute an illusory experience is importantly limited, for perceptions of an appropriately oriented faux penny *are* illusory, even in the narrow sense. Thus on a charitable reading the point of the tilted penny case has never been primarily to suggest that we typically misperceive tilted pennies, it has been to suggest that perceptual ambiguities can yield illusions, even narrowly construed. On the present account we typically avoid misperceiving tilted pennies because of evolutionary and earlier life experience, factors that have been adjusted by reference to our environment and hence have absorbed relevant contingencies like the Euclidean character of local space, the relative absence of elliptical objects (when considering evolutionary and life learning) and the roundness of pennies (when considering life learning). As we will see (Sections 6–7), the above errors have a cost: the familiar but mistaken idea that perceptions of tilted pennies are illusory was incorporated into some influential formulations of the argument from illusion, thus weakening them; but the idea that there is no illusory element to perceptions indistinguishable from those of tilted pennies (e.g. suitable perceptions of faux pennies) has been ignored by Smith and Schellenberg, and weakens their critiques of indirect realism.[18]

5   REVISITING ACQUAINTANCE

Recall that when I mistakenly perceive a blue car to be purple, the indirect realist holds that in making this perceptual error one is acquainted with something other than or in addition to the blue of the car, say a sense-datum that is or at least seems purple – call it a purple sense-datum.[19] By contrast the intentionalist asserts that, if anything, one is acquainted with the blue of the car, but one is experiencing it to be purple because one is merely representing purple, or is in an unsuccessful purple-representing state. In illusory experiences, when considering the mistaken aspect of the experience, is there successful or unsuccessful reference?

---

[18] My critique of Smith appears in Brown, under review.

[19] It is open to the sense-datum theorist to hold that sense-data are coloured and thus that the purple sense-datum *is* purple, but also for her to hold that sense-data are not coloured and thus that the purple sense-datum seems but is not purple. Nothing I say hinges on this dispute: the acquaintance dispute is about whether or not the sense-datum exists (and is an object of acquaintance), *not* about whether or not the sense-datum is coloured. This is one respect in which considerations regarding linguistic, instead of perceptual, reference can lead us astray.

The epistemology of acquaintance generates difficulties: How do I know if, during some illusory experience, the erroneous aspect of that experience consists in me experiencing an instantiated property, or me merely representing an uninstantiated one? Given space constraints I cannot delve into the plethora of ways this epistemic rut has become so difficult to maneuver out of. I will simply assert that in my estimation the rut is so deep that many now feel that the best way 'forward' is to search for other reasons to prefer an acquaintance or non-acquaintance approach to perceptual awareness more generally and import that result – should it ever emerge – into one's account of illusion.[20] This is a mistake.

5.1 In support of intentionalism

A sample case which arguably does not favour an acquaintance interpretation is the floating-vase illusion, but we must take care in our deliberations. When one sees the black ellipse in front of the vase as a shadow one is not seeing an objective shadow, for there is no shadow in that portion of the scene.[21] But this alone does nothing to decide the case, for we as yet have not explained how this misperception has arisen and in turn why its presence does not motivate the existence of a subjective intermediary, a sense-datum that at least seems black, elliptical, and shadowlike. We begin to see why this case does not support an acquaintance interpretation when we recognize that the black ellipse itself can be seen as a shadow cast by the vase *or* as a region of paint in front of the vase. That portion of the stimulus is ambiguous and hence susceptible to multiple interpretations by our vision systems. Similarly, the presented information regarding the spatial distance between the front of the ellipse and the front edge of the base of the vase is ambiguous regarding whether or not there is a vertical component. If there is a vertical component, then the vase is floating, and if there is not, then the vase is on the ground. But the stimulus itself is consistent with both interpretations.

Thus, key aspects of this stimulus are objectively and inherently ambiguous: the way this scene is presenting itself to the agent is perceptually indistinguishable from the way a very different scene (e.g. one with an actual floating vase) presents itself to the agent. Let us accept this. One's vision system prompts one to see the stimulus as a determinate state

---

[20] Some, like disjunctivists, take a rather different approach.
[21] Note that I am not treating 'see' as a guaranteed success term.

of affairs, in this case as a floating vase, and one subsequently can undo this prompting and see the stimulus as a determinate non-floating vase behind a black blob of paint. The act of turning this objectively ambiguous stimulus or given into a perceived determinate state of affairs can be aptly deemed *disambiguating-by-representing*.[22] Importantly, the act is something done *to* what is given in perception, as what is given is *itself* indeterminate and perceivably so, and it is the heart of the misperception giving rise to this illusion, for it is a kind of act whose result can be correct, or can be incorrect. The explanatory challenge is to understand how the ambiguous stimulus with which one is engaged can be, through our representational perceptual capacities, not only experienced as determinate, but experienced incorrectly.

When one erroneously sees the floating-vase stimulus as a floating vase, the familiar indirect-realist strategy has been to postulate a determinate given, sense-data that have the 'floating' property (whatever that means). This is just an instance of their general acquaintance approach to illusions, hallucinations, and misperceptions generally, namely to hold that when one sees something to be x that is in fact not x (e.g. sees the vase stimulus as containing a floating vase), one must thereby be acquainted with, or in a state referring to, something that *is x*. Since the relevant parts of the external world are not that way, what is must be something else, call it a sense-datum. The trouble is that this strategy loses what is most central and interesting about the vase case and others like it. One is in a perceptual state that is directed toward, if not referring to, the objects (/properties/facts) in this scene (i.e. the actual vase, the actual blob of paint, etc.). However, one's understanding of what those objects are must be gleaned from the way those objects are presenting themselves to one in this circumstance, and in this circumstance those presentations do not uniquely determine those objects, they instead determine a collection of candidate disambiguations. So one has to disambiguate what is given and see the stimulus as a determinate state of affairs. But, importantly, even once this is achieved, once one, say, sees the stimulus as a floating vase, one can still recognize that the stimulus is consistent with a very different determinate state of affairs.

---

[22] Instead of 'representing' one could use 'interpreting', 'conceiving', and so on. All choices come with costs and benefits, and it is my hope that as little of the baggage owned by 'representing' is read into my account as possible. For example, as earlier mentioned I wish to avoid discussion of whether or not the act of disambiguating-by-representing can involve the concepts of higher cognition.

Thus the inherent ambiguity of the stimulus cannot be removed from one's account of this illusion, and postulating that what is perceptually given is a determinate sense-datum with the floating property seems to do exactly that, thereby losing what is at the heart of the case.

Alternatively, the indirect realist could postulate sense-data that generate an ambiguous stimulus, but doing so does not yet introduce disambiguating-by-representing into our account of the illusion, and doing so by reference to these sense-data gets us no further along in understanding how disambiguating-by-representing works generally, let alone for sense-data or for the external world specifically. There is furthermore no antecedent reason to suppose, indeed we should resist the suggestion, that disambiguating-by-representing can *only* be done to something subjective. It is true that *in the end* the best explanation of this illusion may involve postulating an additional, subjective perceptual given, other than the vase-illusion stimulus – something like a sense-datum. But at first (and second) glance it is entirely unclear why this additional subjective object would aid in one's explanation. In short the indirect realist approach affords no insight into this illusion.

The intentionalist approach (as I construe it) is by contrast centered precisely on the mark. It begins by accepting that the vase stimulus is itself objectively ambiguous, and that it, if anything, is the object of acquaintance of one's floating-vase perceptions. However, when one sees that stimulus as a floating vase one is taking that stimulus and representing it in an incorrect way.[23] One is not constructing a subjective object of acquaintance with the incorrectly ascribed properties, one is taking an objective object that is from one's perspective indeterminate, and incorrectly (or merely) representing it. Thus, the only object of the perception is the stimulus itself.

To summarize, an approach that postulates a determinate given to explain this misperception might afford a means of explaining disambiguating-by-representing, but does so by leaving behind the core fact

---

[23] In this respect Brewer's (2007a, 2007b) relationalist account of illusions is for our purposes no different from the intentionalist's. Both can agree that there is an existing object that is the object of this perceptual state. They disagree on how this object *gets to be* the perceptual object: for the intentionalist it is through being represented by one's current state and for Brewer it is simply by one's being 'perceptually open to the world'. Once this difference is registered, their accounts of illusion are not obviously different, for both agree that in illusory perceptions one is conceiving of, or representing, the perceptual object to be some way that it is not. Whatever difference there is between them hinges on how this 'conceiving of' or 'representing' gets fleshed out, something Brewer does not say much about, and intentionalists typically say what I've been presuming.

that what is given is inherently ambiguous. By contrast one that postulates an inherently ambiguous but subjective given leaves the inherent
ambiguity of the given in tact, but affords no obvious insight into how
disambiguating-by-representing might work, let alone why it cannot simply work on the objectively ambiguous. The correct approach takes the
objectively ambiguous vase stimulus as the heart of one's account of the
perceptual objects of this illusion, and seeks to explain the misperception
by explicating how disambiguating-by-representing can be applied by our
minds to such a thing. This is how I understand the intentionalist strategy,
and it involves utilizing the failed referential capacity of mere representation. To be sure I do not believe that this explanation of the vase illusion is complete, or even near complete, for there is as yet no explication
of disambiguating-by-representing.[24] My point is that it is correctly oriented, on the right path, whereas the indirect realist explanation seems to
be on the wrong one. The conclusion that this correct path points toward
is that the vase illusion, and perhaps all illusions due to ambiguities, do
not give evidence for sense-data. Fortunately for the indirect realist, other
illusions do.

5.2   In support of sense-datum theory

Consider the Hermann grid illusion.[25] When one engages with such a
grid one sees in front of one, at many intersections of the grid on the
page, items one is apt to describe as 'small black dots'.[26] Misperception
is occurring because there objectively are no dots at these intersections
or anything mind independent – like peculiar light reflectances – which

---

[24] One can see Macpherson's (2006) argument from ambiguity against intentionalism as fitting here,
as an argument for why in some alternative cases (e.g. the square-diamond case) what I've called
disambiguating-by-representing cannot be glossed in the manner dictated by intentionalism. I
regret not being able to remark on her contribution here, except to say that the vase case is, from
the perspective of her discussion, much more like the duck–rabbit than the square diamond. The
reason is that the floating and non-floating disambiguations of the vase stimulus have distinct
objective characteristics, much like ducks and rabbits do, and unlike (in Macpherson's view) the
square diamond does. For this reason she is likely to endorse my claim that the vase case can be
given a reasonable explanation by intentionalism. It is also worth noting that her opposition to
intentionalism is not married to a defense of indirect realism specifically. See Raftopoulos (2011) for
a response to Macpherson's argument.
[25] Other illusions one could straightforwardly apply to following analysis to include spreading illusions, the watercolour illusion, etc.
[26] In case there is doubt, I think that it is safe to say that one sees these dots in both the doxastic and
phenomenal sense, and that the grid looks to have these dots in both senses (see e.g. Dretske 1995).
Focusing on differences such as this is important in some contexts, but not in this one.

could answer to 'black dots'. To be sure the dots have strange properties, for example when one focuses on a particular intersection no dot is there seen, but a dot does appear there when one shifts one's focus to an adjacent area of the grid. The strangeness of these experienced dots is not at issue. And, as per our above discussion (see note 19), the issue is also *not* whether or not it is correct to call the dots 'black'. Issues about linguistic reference are informed by, but not constrained to, first-person experience and report. The *core* issue is whether the mistaken aspect of the experience is best described as one involving the perceiver being aware of an instantiation of black (which she is inclined to refer to as a 'black dot'), as the acquaintance theorist would hold; or whether the experience is best described as one *not* involving the perceiver being aware of an instantiation of black, but instead as seeing in a particular 'black-dot-ish' way or being in a perceptual state that merely represents black dots (that in fact do not exist), as the non-acquaintance theorists (adverbialists and intentionalists, respectively) would hold. When engaging the Hermann grid, are the experienced black dots best construed as experiences of instantiations of black, or as mere representations of blackness?

The fact that there is nothing in the objective stimulus that can help explain the illusory experience – an *exact opposite* to the vase case – is important because it means that the bulk of our explanation should be subjective: one's mind is doing something peculiar, it is adding something to the experience that is in no way part of, suggested by, or even consistent with the stimulus itself. The curious thing is that one's mind is not merely adding something to the experience, it is *adding features to the object of the perception*, the little black dots you experience to be at the grid intersections.[27] The existential import of such descriptions often make intentionalists cringe, but they are not the indirect realist's construction. (1) If you ask a perceiver whether or not she actually experiences little black dots at various grid intersections she answers affirmatively. (2) If you ask her whether or not she can understand those little black dots as non-existing or uninstantiated entities, as things that she is *merely* representing, she does not answer affirmatively, but instead asks what you mean. (3) If you ask a perceiver to draw or paint a snapshot of what she sees in the vase case

---

[27] This is why the projectivist variety of indirect realism is needed for the present work; these purportedly subjective perceptual objects are experienced as being in the world outside oneself, on the grid's surface. This is also why the analysis given by the typical qualia realist (e.g. Block and Stoljar), according to which the mind is adding non-intentional qualitative aspects to the experience, is inadequate. Unfortunately, I cannot discuss the latter here.

(i.e. the scene as it is viewed from her perspective), the drawing would be the same in the floating and non-floating experiences of the stimulus. A drawing of a snapshot of what she sees in the grid case would consist of a grid with little black dots at various intersections. In the latter something is being added to what one is experiencing, whereas in the former one is merely interpreting what one is experiencing in a certain way. (4) If you ask her if she sees white at the relevant grid intersections she will say 'no', that even though she knows there is white there, the black dots are in the way, they are intervening between her and the objective colour of those intersections. This *is* the phenomenon that needs explanation, that from one's perspective one is experiencing an instantiated set of features – black dots – no members of which match up with anything in the objective stimulus. And the most reasonable starting place is to accept that these are instantiated perceived features that are subjective but projected, to accept the orientation afforded by sense-datum theory. Indeed the whole reason the grid case is of interest is because it gives evidence for our minds' capacity to add features that have no trace in the objective objects of perception.

The intentionalist will try to preserve each (or at least most) of these descriptions without committing herself to the existence of instantiated black dots, so that when pressed she can always reply, 'but one is not experiencing instantiations of black, one is instead merely representing them, or having representations as of them'. Consider some possible replies. (1 and 3) She says the dots actually exist/she draws dots in her depiction of what she sees, but she is mistaken: they only *seem* to her to exist. (2) She does not understand why I would try to convince her that the black is not instantiated, or even what that might mean (for her experience), but the correct perceptual theory must be learned, not gleaned from a subject's untrained response. (4) She cannot see the whiteness at the grid intersections because she is not representing it but is instead representing blackness, not because she sees some instantiated property that is in the way. This overall strategy is in effect to *deny the very phenomenon that is the illusion*. In this sense an error theory about phenomenology and first-person descriptions is required by the intentionalist approach to this and like illusions. Intentionalists are aware of this, but see this skepticism about phenomenology as a minor tax for upholding their view. This explanation is inadequate.

The key question is to ask whether or not the tools of one approach, those of the sense-datum theorist or the intentionalist, better explain the subtleties of the case. The phenomenology of grid perceptions, or the grid stimulus as it is experienced from the first-person perspective, consists of an actual grid with instantiated black dots at various intersections – call this the *agent's data*. These dots are not objective and so are most likely subjective features added, for whatever reason, by our vision systems. The acquaintance interpretation gives a straightforward, natural explanation of this illusion. The direct realist explanation begins by denying the veridicality of the *agent's data*: the agent doesn't experience instantiated dots; it only *seems* to her as though she does. What is happening, on this view, is the agent is merely representing these dots, she is in a dot-representing state that is in fact representing no dots. But this only generates a new puzzle: How does the idea of mere representation explain the *agent's data*? The direct realist is arguing that what the agent is actually aware of is opaque to her, that the *agent's data* are, for some reason, fundamentally erroneous.[28] The idea of mere representation does nothing to assist in our understanding of this. Let me briefly elaborate.

The idea of mere representation (i.e. representation with whole or partial reference failure), whose home is secure in non-perceptual thought and various other propositional attitudes, was never supposed to immediately and fully explain perceptual illusion (or hallucination). It was supposed to be a model that could be developed to do so. And I think it can, but only to a limited extent. The underlying issue is the extent to which being in a particular mental state gives one evidence to believe in the existence of the objects of that state. Call this the *existential commitment* of a state. Setting aside issues regarding purportedly necessary objects like numbers, and *pace* Meinong, representation in thought comes with virtually no existential commitment to its objects. Thinking about vampires gives me no reason to think that they exist, it in no straightforward way makes it even seem like they exist. This is so in part because thinking about them as existing or not is arguably equally easy, all else being equal.[29] One can (though I

---

[28] I take it as given that this challenge is not met by speckled hens or finger counting (see Dretske 1995, chapter 5 for the latter). Given a large Hermann grid and a long period of time I may still be wrong about how many dots I see. That is quite different from saying that I can be wrong about whether or not I see instantiated black dots. And *that* is quite different from saying that it *is* wrong that I actually see instantiated black dots. This last judgement must overcome the existing evidence to the contrary, and it must do so through an argument, not merely through an intentionalist theory that entails it.

[29] One (defeasible) reason some concerned with the a priori believe in the existence of numbers is because of the necessity of arithmetical truths, a necessity partly grounded in the fact that, as Frege says, if one denies the laws of arithmetic, 'to think at all seems no longer possible' (*Foundations of Arithmetic*, §14).

would not) argue that representation in fears and desires can bring slightly more existential commitment. I am locking the door because my fears of being bitten by a vampire make it seem like there are such things, even though I have ample alternative evidence to the contrary. On such a view 'seem' is read as 'give evidence for the existence of'.

Regardless, perception is the *opposite* of thoughts, fears, and desires on this issue, for it is generally held to involve a considerable degree of existential commitment. Unless one is a skeptic about the external world, being in perceptual states is generally taken to give one reasonable evidence for the existence of the objects of those states. We thus have a principled reason to *not* extend this reasoning (about thoughts, fears, and desires) to perceptual states, and to instead, with perceptual states, begin at the opposite end of the existential commitment dimension. If the intentionalist wishes to inject into perceptual states the kind of skepticism needed to jettison the first-person evidence for sense-data in grid and like illusions, she needs more than a loose analogy with non-perceptual states and their use of mere representation.[30]

This doesn't prevent us from giving alternative arguments to assess the existential commitment of troublesome perceptual cases. On my account, the vase case brings with it about as much existential commitment as does the fear that prompts one to lock the door (which is very little): it seems like that's a floating vase, even though (without moving) I can now see it as a vase sitting on the ground. We should seek an explanation, like the intentionalist one, that preserves the inherent ambiguity of the stimulus, and one that does not add things to our ontology that are not suggested by the case. The grid case, however, involves *much* greater existential commitment to what objectively is not present in the stimulus: there is simply no other way to see the stimulus other than as a grid with black dots at various intersections. Telling me that these dots are not actually instantiated (by my mind and projected onto what I experience), that instead I am merely representing them, does nothing to illuminate why I am wrong, it simply tells me that I am. We here see the explanatory limits of the mere representation tool, and a distinct advantage of sense-datum theory.

It may help to think of it in this way. We shouldn't feel the need to postulate something that instantiates the floating property, because we can

---

[30] Thus, whatever the merits of Harman's ([1990] 1997) Ponce de Leon argument, it cannot be expected to apply to perceptual illusions en masse.

see the stimulus itself as something that need not have that property. The same is not true of the dots: when I see them, I can't help but see them. Particularly once we consider the *possibility* that the blackness is subjective, it is as easy to doubt that I am seeing several instances of black as it is to doubt that I am seeing the grid. And the mere representation tool gives no insight into how to motivate the level of skepticism needed to uphold the intentionalist analysis.

My general point therefore is twofold. Neither the sense-datum theorist nor the intentionalist should be trying to fit phenomena into his view, he should be trying to take from phenomena data that support some view or other. I believe I have shown that, at least to a greater extent than is typically presumed, this is possible. Secondly, some illusions support the intentionalist approach, and others the sense-datum approach.[31] It is from here that we move forward and assess the impact illusion has on philosophical argumentation about perceptual theory. Let me give you one example of how this might work.

6   WHERE'S THE FIRE? (THE ARGUMENT FROM ILLUSION)

There is not any one argument that owns the term 'argument from illusion', but the broad aim of all such arguments is to persuade us that:

(1) illusory perceptions involve awareness of sense-data, and thus
(2) all perceptions plausibly involve awareness of sense-data (i.e. direct realism is false).[32]

The move from (1) to (2) is non-trivial, and is of little interest should (1) be unsustainable. To be assessed (1) must be supplemented with a conception of illusion (e.g. narrow or wide) and with guidance regarding how many illusions plausibly involve awareness of sense-data. Thus an unrestricted reading of (1) presumes a wide conception of illusion and that all illusions involve awareness of sense-data, whereas a restricted one might presume a narrow conception and that only some illusions involve awareness of sense-data. This flexibility naturally interplays with the inference from (1) to (2). If one accepts an unrestricted reading of (1), then all misperceptions involve illusion, and since misperceptions

---

[31] There are many other illusions that are subject to one of these two analyses, and many others that are not. Illusions that arguably require additional arguments to incorporate them into this dispute include the twisted-cord illusions, various colour, stereo, and motion illusions, etc.

[32] Smith 2002 and Gupta 2006 are two recent and very worthwhile discussions of the topic.

can be had with any perceivable object and in virtually all perceptual circumstances (that allow at least some contingencies) the leap to the claim that (2) all perceptions involve awareness of sense-data is not far. By contrast on a restricted reading of (1) only a small subset of misperceptions involve illusion, those containing sensory trickery, and perhaps only some of those require us to postulate a perceiver's awareness of sense-data. From here the leap to (2) is substantial, for admitting that some instances of sensory trickery require the presence of sense-data does not straightforwardly generalize to the conclusion that all perceptions do too. As the reader may already have guessed, on my view the latter option is preferred.

The core issue underlying our assessment of (1) is the extent to which acquaintance interpretations should be given to perceptual illusions. If an acquaintance interpretation is warranted, then the perceiver's misperception entails the awareness of something which has the erroneously attributed features. Given that the perceived part of the external world does not have anything to answer this call, the item of awareness cannot be in the external world, making it most plausibly part of the internal or subjective world of the perceiver (i.e. sense-data). I have argued that there are illusions that should be given an acquaintance interpretation, and thus that during these illusory perceptions agents are aware of sense-data. It does not follow that all narrow illusions should be understood similarly (e.g. the vase illusion should not), or that wide illusions should be (e.g. randomly mistaking my sister's for my mother's voice). In other words, I support a restricted reading of (1), a starting point that I hope is amenable to enough readers, and that will help us to see one place in perception where we plausibly *begin* to lose grip on the external world. However, getting from here to (2) – to a conclusion about the *extent* to which we have lost grip on reality – requires some ingenuity. That matter I regrettably leave to another time, for enough has been given to expose the shortcomings of an important direct realist trend.[33]

---

[33] This strategy is quite different from the more familiar one that employs a broad reading of (1). In outline, the latter strategy holds that: for any given perceptual object there is a feature that can be misperceived; misperceptions constitute perceptual illusions (i.e. illusions are wide); and all illusions should be given an acquaintance interpretation. The first claim is doubtless true, and the second, while subject to definitional disputes, is not the philosophical heart of the approach. The heart is the last step, which in another incarnation is the infamous sense-datum inference: 'whenever something perceptually appears to have a feature when it actually does not, we are aware of something that does actually possess that feature' (Smith 2002, 25). See also Robinson's 'phenomenological principle' (1994, 32). (It is not clear to me that most indirect realists have explicitly

7 THE ARGUMENTS FROM PERCEPTUAL RELATIVITY
AND FROM ILLUSION

The key difference between the argument from perceptual relativity and
the one from illusion is that the former asserts that mind-dependent perceptual objects are needed to account for the varying *ways* that objects
perceptually look in differing circumstances, and the latter asserts that
mind-independent objects are needed to account for *misperception* (or at
least the special kind peculiar to illusions). The two collapse if the ways
in need of explanation are presumed misperceptions and all misperceptions are presumed illusory (i.e. illusions are wide). Both assumptions
were arguably widely held in much early modern philosophy, but I have
already explained why in our current climate both assumptions should be
rejected. What remains is to appreciate the difference between the argument from illusion and the one from relativity that emerges, and in particular the additional demands of the former.

There is a sense in which the shape of Russell's table[34] looks different as
one moves around it, that there are various ways the table's shape looks.
One could hold that these ways can only be explained by positing mind-
dependent perceptual objects, and hence use this data – as Russell perhaps
does – to argue for indirect realism. But why would we ever expect the
way an object's shape looks to be the same in all or even most circumstances? The answer is that in perception we should not expect this, and
it is not true.[35] This does not remove the burden of explaining how these
differing looks arise, but it does open the door for doing so by appeal
to mind-independent relative features. The trick to making this response
work is to give a coherent account of these relative properties, and that
is one aim of Schellenberg's (2008) account of SD (situation-dependent)
properties.[36]

defended this idea, but I can appreciate how its seed can be drawn from their views.) As these
authors argue, there are various difficulties with the sense-datum inference and various ways to
modify it to make it more palatable (see Smith 2002, chapter 1; Robinson 1994). With respect to
this issue my claim is that the sense-datum inference is too coarse to be of much use. I unfortunately cannot give a full discussion of this matter.
[34] See the opening pages of *Problems of Philosophy* ([1912] 1959).
[35] This point was emphasized by Dawes Hicks (1912) and Demopoulos (2003). The issue, however,
deserves closer scrutiny than space permits.
[36] It is also an aim of Dawes Hicks' (1912, 1913/14) contributions, Dummett's ([1979] 1993) absolute
and relative descriptions of the world, Demopoulos' (2003) use of Dummett's distinction, Maund's
(forthcoming) account of constancy, and so on. I focus on Schellenberg because of the relative ease
with which her work fits into the present discussion.

Schellenberg's core distinction is between SD and intrinsic properties of perceived things. Intrinsic properties are 'the properties that an object has regardless of the situational features … [they do] not depend on the object's relations to other individuals distinct from itself' (2008, 55). By contrast situational features are 'the features of the environment that determine the way an object is presented' and include for example 'lighting conditions and the subject's location in relation to perceived objects' (56). SD properties are the way an object is presented in some perceptual circumstance, and are a function of its intrinsic features and the prevailing situational ones. Her thesis is that in perception 'the way an object is presented is best understood in terms of external, mind-independent, but situation-dependent properties that the object has given its intrinsic properties and the situational features' (56–57). And, more boldly, she closes her article with the following: 'If one recognizes situation-dependent properties, no appeal to mind-dependent properties is necessary to explain how it can be that there is a way that objects look that is not accounted for by representing their intrinsic properties. If this is the reason for introducing mind-dependent properties (or objects), then one might as well drop them once situation-dependent properties are acknowledged' (84). I take the implication to be that this last conditional's antecedent is plausible.

All should agree that 'ways of appearing' is *one* reason indirect realists postulated mind-dependent perceptual objects. But is it the chief reason? Schellenberg notes some others: 'One is that [mind-dependent perceptual objects] make an experience an experience. A second is to account for the possibility that hallucinatory and veridical experiences are phenomenologically indistinguishable. A third is to account for the possibility of spectrum inversion' (71). Interestingly, illusions are not on this list. She seems to take her argument for SD properties and for the non-illusory status of some familiar examples of illusions (e.g. on pp. 74–75 she discusses Peacocke's trees and the half-submerged stick) to undermine any residual argument from illusion. If this is so,[37] then she has erred, largely for the reasons stated earlier, and it is worth rearticulating this point with reference to Schellenberg's view.

When we say that the table looks trapezoidal from here we can add that it also looks rectangular from here. A contradiction can be avoided for example when we interpret the first description as referring to a relative or

---

[37] One could instead speculate that she means to include illusions with hallucinations. However, in the article she cites Smith's (2002) book, which is painfully clear about why the argument from illusion is crucially different from the argument from hallucination.

SD shape property and the second to an absolute or intrinsic one. Only if relative shape properties require the existence of mind-dependent perceptual objects (and they do not) does this give evidence for indirect realism. This is Schellenberg's (and Dawes Hicks', Dummett's, Demopoulos', etc.) point. However, another argument for indirect realism can still be marshalled, for once we accept that we can simultaneously perceive relative and intrinsic shape properties of the table, the stimulus it constitutes becomes ambiguous: it is consistent with there being a trapezoidal table that one is viewing head-on, or a rectangular table one is viewing from an angle.[38] Our vision system does not give these disambiguations equal weight. Using background assumptions and additional cues in the scene it (by hypothesis) correctly gives the highest weighting to the 'rectangular and viewed at an angle' disambiguation, prompting us to reflexively see it as rectangular. But as earlier noted this same set of background assumptions and cues can be used to generate a stimulus that prompts our vision systems to give the wrong disambiguation the highest weighting and thereby results in perceptual illusion. A trapezoidal table viewed head-on would, with some additional controlling features, suffice for this purpose. Two points need emphasis.

Schellenberg states that '[i]f one recognizes [SD] properties, then many cases of perceptual experiences that in the philosophical literature are typically categorized as illusory or as misrepresentations will turn out to be accurate perceptions – at least with regard to their representations of [SD] properties' (74–75). This is correct, but it is so because of our adeptness at perceptual disambiguation, the explanation of which requires much more than the admission of SD and intrinsic properties into our theory. It additionally requires some mechanism for disambiguation which I suspect must be mental or cognitive (and in any case in our situation *is* mental), some specification of the preferences subpersonally associated with each disambiguation, and some account of why these preferences are present. SD properties on their own represent one small (albeit important) step in this explanation. In other words, if one recognizes SD properties, many purported illusions 'turn out' to be accurate perceptions, but the reason they turn out that way has to do with much more than the involvement of SD properties.

---

[38] This is not to suggest that simultaneously perceiving some absolute and relative property of an object will always or even usually foster perceptual ambiguity. It is to say that in this case the simultaneous perception of such properties is doing so. I regret not being able to analyze the more general issue.

Secondly, and more pointedly, admitting relative shape properties into one's ontology does not remove the threat of illusion, it only displaces it, and once illusions are present a wholly different problem emerges: we need an explanation of the contained *misperceptions* that does not postulate mind-dependent perceptual objects, for otherwise we are again confronted by indirect realism. SD properties do nothing to aid with this struggle. Indeed the first point lays the groundwork for the second: it is because the admission of SD properties does not explain why familiar examples of purported illusions are not illusions that SD properties give no insight into how we should understand the misperceptions inherent in illusions. Unfortunately, Schellenberg shows no awareness of this point, no awareness of the difference between the argument from perceptual relativity and the argument from illusion. In this, however, she is not alone: many indirect realists, like Russell, have failed to clearly make the distinction.

8   CONCLUSION

Understanding why some illusions do and some do not support the existence of sense-data is a non-trivial task. In this work the core issue is whether the misperceived aspect of an illusory experience should be given an acquaintance or merely representational interpretation, whether or not it should be understood as involving successful perceptual reference. To elucidate the latter I focused on illusions due to ambiguous stimuli that trick us by preying on the interpretive biases in our vision systems. In these cases what is given to a perceiver is inherently ambiguous and the vision system's task is to select the correct disambiguation. The difficulty is that these stimuli prompt the vision system to select an incorrect one, thus generating an illusory experience. In such an experience the perceptual *object* is most plausibly taken to be the ambiguous stimulus, for that is what the perceiver sees to be some way or other. And even when we see it determinately (whether correctly or not), we can still see its inherent ambiguity, at minimum by subsequently seeing it in some other determinate way. The heart of this illusion is thus the act of disambiguating-by-representing an objectively and perceivably ambiguous stimulus and understanding why an incorrect disambiguation is being favoured by one's vision system. The idea of mere representation – the intentionalist's tool – has elegant application here: when the vision system selects an incorrect disambiguation, it is not creating determinate objects of acquaintance with the incorrectly attributed properties, it is merely representing the ambiguously presented object of acquaintance erroneously. Introducing unambiguous subjective perceptual objects to explain the illusion is not only misdirected, it leaves the core puzzle untouched.

By contrast some illusions are only obscured by attempts to understand them in this way. In the Hermann grid, the features relevant to the mistaken aspect of the illusory experience – the black dots – have no objective trace, and when I see them I cannot help but see them as instances of black. The level of first-person experiential doubt needed to deny this is much greater than that supported by thoughts about non-existent things, intense desires/fears about non-existent things, the vase illusion, and so on. The intentionalist explanation for the opaqueness of the nature of the experience to the perceiver consists of an application of the mere representation tool: the perceiver is not aware of instances of black but is merely representing them. The difficulty is that this tool offers no insight into why the true nature of the experience is opaque to the perceiver or why the perceiver phenomenologically experiences what she does. Mere representation has worthwhile applications elsewhere but falls flat as an explanatory tool here. By contrast the black dots are naturally viewed as instantiated features subjectively added to the object of one's perception – they are sense-data.

What emerges from the thesis that some illusions do and some do not support the existence of sense-data is that the argument from illusion is far from compelling, removing much of the force behind calls to reject sense-data for fear that doing so will lead to indirect realism generally. In addition the framework used to articulate these ideas exposes a fundamental difference between arguments from illusion and those from perceptual relativity, and explains why defeating the latter has little effect on the former.

REFERENCES

Armstrong, D. (1961) *Perception and the Physical World*. New York: Routledge.
Brewer, B. (2007a) Perception and its objects. *Philosophical Studies* 132: 87–97.
  (2007b) How to account for illusion. In F. Macpherson and A. Haddock (eds.) *Disjunctivism*. Oxford University Press.
Brown, D. (2010) Locating projectivism in intentionalism debates. *Philosophical Studies* 148: 69–78.
  (Forthcoming) The transparent projectivist. In F. Macpherson, F. Dorsch, and M. Nida-Rümelin (eds.) *Phenomenal Presence*. Oxford University Press.

(Under review) Getting acquainted with perceptual constancy in early analytic philosophy. In M. Dumitru (ed.) *The Actuality of Early Analytic Philosophy* [working title].

Byrne, A. (2001) Intentionalism defended. *Philosophical Review* 110, no. 2: 199–240.

Crane (2006) Is there a perceptual relation? In T. Gendler and J. Hawthorne (eds.) *Perceptual Experience*. Oxford University Press, pp. 126–146.

Dawes Hicks, G. (1912) The nature of sense-data. *Mind* 21, no. 83: 399–409.

(1913/14) Appearances and real existence. *Proceedings of the Aristotelian Society*, n.s., 14: 1–48.

Demopoulos, W. (2003) Russell's structuralism and the absolute description of the world. In N. Griffin (ed.) *Cambridge Companion to Bertrand Russell*. Cambridge University Press, pp. 392–419.

Dretske, F. (1995) *Naturalizing the Mind*. Cambridge, MA: MIT Press.

Dummett, M. ([1979] 1993) Common sense and physics. Reprinted in *The Seas of Language*. Oxford University Press, pp. 376–410.

Gupta, A. (2006) *Empiricism and Experience*. New York: Oxford University Press.

Harman, G. ([1990] 1997) The intrinsic quality of experience. Reprinted in Ned Joel Block, Owen J. Flanagan, and Güven Güzeldere (eds.) *The Nature of Consciousness*. Cambridge, MA: MIT Press, pp. 663–676. Originally published in J. Tomberlin (ed.) *Action Theory and Philosophy of Mind*. Philosophical Perspectives, vol. IV. Atascadero, CA: Ridgeview Publishing Co., 1990, pp. 31–52.

Hellie, B. (2006) Beyond phenomenal naiveté. *Philosophers' Imprint* 6, no. 2: 1–24.

Hilbert, D. (2004) Hallucination, sense-data and direct realism. *Philosophical Studies* 120: 185–191.

Hochberg, J. and Peterson, M. A. (1987) Piecemeal organization and cognitive components in object perception. *Journal of Experimental Psychology: General* 116: 370–380.

Macpherson, F. (2006) Ambiguous figures and the content of experience. *Noûs* 40: 82–117.

Maund, B. (2003) *Perception*. Montreal, QC: McGill-Queen's Press.

(Forthcoming) Perceptual constancies: Illusions and veridicality. In C. Calabi and K. Mulligan (eds.) *Essays on the Illusions of Outer and Inner Perception*. Cambridge, MA: MIT Press.

Raftopoulos, A. (2009) *Cognition and Perception: How Do Psychology and Neuroscience Inform Philosophy?* London: MIT Press.

(2011) Ambiguous figures and representationalism. *Synthese* 181, no. 3: 489–514; doi: 10.1007/s11229-010-9743-1.

Robinson, H. (1994) *Perception*. New York: Routledge.

Russell, B. ([1912] 1959) *Problems of Philosophy*. Oxford University Press.

(1913) The nature of sense-data – A reply to Dr. Dawes Hicks. *Mind* 22: 76–81.

Schellenberg, S. (2008) The situation-dependency of perception. *Journal of Philosophy* 105: 55–84.

Seckel, A. (2003) *Incredible Visual Illusions*. London: Arcturus.

Smith, A. D. (2002) *The Problem of Perception*. Cambridge, MA: Harvard University Press.

Spelke, E. S. (1990) Principles of object perception. *Cognitive Science* 14: 29–56.

Tye, M. (2000) *Consciousness, Color, and Content*. Cambridge, MA: MIT Press.

Wandell, B. A. ([1989] 1997) Color constancy and the natural image. In A. Byrne and D. Hilbert (eds.) *Readings on Colour*, vol. ii: *The Science of Colour*. Cambridge, MA: MIT Press, pp. 161–176. Originally published in *Physica Scripta* 39: 187–192.

CHAPTER 5

Perceiving the intended model

John Campbell

I TRADING ON IDENTITY

My concern in this chapter is with our knowledge of the references of perceptual demonstratives: terms like 'this' and 'that' used to refer to currently perceived objects, such as a tree or a person. I want to orient the discussion with respect to a basic question about our understanding of arithmetic. It's often said that the first thing one has to grasp about our understanding of arithmetic is that it's natural to think of it as a matter of having a quasi-perceptual grasp of the intended model of arithmetic, and that this natural idea can't be quite right as it stands. The point I want to make in this chapter is that this natural idea, whether it's right or wrong in the arithmetical case, seems illuminating in the case of perceptual demonstratives. After all, here it is not that we have a 'quasi-perceptual' grasp of the intended model; we do literally perceive the intended objects and properties.

Suppose we construct a formal system and give a syntactic definition of a theorem in the system. By Gödel's proof, we will be able to construct a statement U that we can recognize to be true, but that does not meet the syntactic definition of a theorem. In 'The Philosophical Significance of Gödel's Theorem', Michael Dummett remarks:

A common explanation is as follows. Since U is neither provable nor refutable, there must be some models of the system in which it is true and others in which it is false. Since, therefore, U is not true in *all* models of the system, it follows that when we say that we can recognize U as true we must mean 'true in the *intended* model of the system'. We must thus have a quite definite idea of the kind of mathematical structure to which we intend to refer when we speak of the natural numbers; and it is by reference to this intuitive conception that we recognize the statement U to be true. On the other hand, we can never succeed in completely characterizing this intuitive conception by means of any formal system, that is, by any finitely stateable stipulation of the set of statements about natural numbers which we are prepared to assert. (Dummett 1978, 186)

The natural, though not necessarily correct, way of interpreting this talk of a 'quite definite idea', or 'intuitive conception' of the intended model is in quasi-perceptual terms. As Dummett later puts it, this account:

> operates with the notion of a model as if it were something that could be given to us independently of any description: as a kind of intuitive conception which we can survey in its entirety in our mind's eye, even though we can find no description which determines it uniquely. (191)

I'll call this the 'common explanation' of the Gödel proof. Dummett offers a number of criticisms of this natural idea; I'll review them in a moment. Right now, though, I remark that there seem also to be limitations on a purely syntactic characterization of our grasp of perceptual demonstratives. Let me first explain the notion of an inference that 'trades on identity'. Consider the argument:

Hesperus is F
Phosphorus is G
Hence, something is both F and G.

This inference is not valid as it stands. It needs an extra premise, 'Hesperus is Phosphorus', before we have a valid argument. Consider now the argument:

Hesperus is F
Hesperus is G
Hence, something is both F and G.

This argument is valid as it stands. There is no need for an extra premise asserting the identity of Hesperus (as referred to in the first premise) with Hesperus (as referred to in the second premise). If you did suppose that such a premise is needed, and provided it, you would have only begun. For you would now need further premises asserting the identity of all the various Hesperuses referred to in the course of the inference; and no finite, or for that matter, infinite provision of further premises would be enough. We have to acknowledge that the argument is valid as it stands. Rather than depending on an implicit identity premise connecting the terms of the first two premises, it simply 'trades on' the identity of reference of those terms. Consider now the following inference, (A). Each use of 'that' is singling out a wasp, in a garden containing many wasps:

A. That is buzzing
   That is a long way from home
   Hence, something that is buzzing is a long way from home.

Does this inference need to be supplemented by a premise asserting identity? Or is it valid as it stands, legitimately trading on identity? My first point is that this is not a matter that can be settled by the syntax of the inference alone. Let us suppose that (A) is not valid as it stands. Perhaps both demonstrative uses of 'that' do indeed refer to the same thing, but that is consistent with the inference being invalid. Perhaps the two sightings of a wasp underpinning the two demonstratives were incidents in the observation of a cloud of wasps making it impossible for one to keep track of the thing between sightings. On the other hand, we could have a valid inference, in which there is manifestly just one thing being referred to:

B. That is buzzing
   That is a long way from home
   Hence, something that is buzzing is a long way from home.

It could be that (B) is valid even though (A) is not. It could be that the demonstratives in (B) are both underpinned by a single sighting of the wasp. Certainly, inferences such as (B) must sometimes be valid as they stand, if inferences exploiting identity that involve perceptual demonstratives are ever to be valid. But the syntax of (A) is exactly the same as the syntax of (B). Whatever the difference between them, it does not seem to be a difference in syntax.

You might argue that the difference between (A) and (B) will in the end turn out to be a syntactic difference, once we look at the details of the mental representations underpinning the uses of the English demonstratives. I will in a moment look at some of the possibilities here. For the moment I remark only that the situation at the level of ordinary English might well be replicated at the level of an underlying system of mental representation. That is, an underlying system of mental representation might well itself use demonstratives that exhibit the same behaviors as ordinary language demonstratives. It might even at the underlying level be possible to formulate arguments (A) and (B) that are syntactically indistinguishable, one of which is invalid and the other valid.

You might say, well the difference between (A) and (B) is a matter of context. My present point is just that the notion of 'context' here is not itself syntactic. Nor is it straightforward to say what sameness of context requires. All the demonstratives have the same references in inferences (A) and (B), for example, so sameness of context cannot itself be a matter merely of sameness of reference.

On the face of it, then, the difference between the invalid argument (A) and the valid argument (B) cannot be determined purely syntactically. I do not want to argue that this comment reflects a level of insight similar to that involved in Gödel's proof that the arithmetical truths cannot be characterized syntactically. But I do want to point out that in its humble way, this point provokes a reaction similar to what Dummett calls the 'common explanation' of Gödel's theorem. That is, it is natural to think that the difference between (A) and (B) has to do with our apprehension of the model intended to be used in interpreting the premises. If the model as intended presents the references of the demonstratives in the premises in just the same ways, then the inference is valid. Otherwise, the inference is not valid.

We could put the point like this. We can define a model-theoretic notion of validity as follows. An argument is valid if, in every admissible model in which the premises all come out true, the conclusion comes out true. To define the notion of an 'admissible model' for a language involving demonstratives, we need to know when it is a condition on an admissible model that it assigns the same reference to a pair of uses of a demonstrative, and when it is admissible for a model to assign different uses to each of a particular pair of uses of a demonstrative. Sometimes it must be admissible for a model to assign different references to a pair of uses of a demonstrative, otherwise inference (A) could not come out as invalid. Sometimes, on the other hand, it must be a condition on the admissibility of a model that it assign the same reference to two different uses of a demonstrative, otherwise inference (B) could not come out as valid. So what we need is a way of saying when an inference is being understood in such a way that any admissible model must assign the same reference to two uses of a demonstrative, and when the inference is being understood in such a way that admissible models may assign different references to the two uses of the demonstrative.

I think that we can state the distinction we need here by using something like Frege's notion of a 'mode of presentation'. We can say that the crucial question is whether the two uses of the demonstrative are being interpreted in terms of the same 'way of being given' the referent. If the two uses of the demonstrative present the referent in the same way, then any admissible model of the language must assign the same referent to the two uses of the demonstrative. If the two uses of the demonstrative use different ways of presenting their referents, then admissible models of the language may assign different references to the two uses. The way in which the intended model is grasped on the basis of perception will be the prototype that defines the class of admissible models.

My point here is that when the grasp of syntax seems inadequate to explain our cognitive achievements, both here and in the Gödel case, it is natural, if not necessarily correct, to appeal to our grasp of an intended model as explaining how we get beyond syntax. I want to set out a bit more below how the idea of a quasi-perceptual grasp of the intended model might apply to the case of perceptual demonstratives, before looking at Dummett's critique of this idea.

2   FUNCTIONALISM

It has often been remarked, as a basic problem in theory of meaning, that the only credible accounts of meaning are truth conditional, but that it is hard to understand how the functional organization of a subject could constitute their grasp of the truth conditions of the statements they make and the thoughts they have. It seems, at first sight at any rate, unquestionable that understanding a proposition is a matter of knowing what has to be the case for it to be true. But it is hard to see how merely being sensitive to certain inputs in one's assessment of a proposition, and tending to produce certain outputs upon accepting a proposition, could constitute knowledge of what it is for the proposition to be true. The underlying idea behind the appeal to our 'quasi-perceptual grasp of the intended model' for discourse involving demonstratives is that there is an epistemic role for consciousness: that sensory experience is not idle in our cognitive contact with the world, not merely an epiphenomenon spun off by our cognitive lives, but something that plays a basic role in our knowledge of our surroundings. This epistemic role for consciousness is what allows us to explain how grasp of truth conditions could transcend one's merely having a particular functional structure.

To see the problem in reconciling a truth-conditional account of meaning with a functionalist account of grasp of meaning, consider a contrast that Dummett draws, in later work, between two types of truth-conditional theory of meaning:

(1) Truth-theoretic accounts of meaning, that merely state, for each sentence of the object language, what it takes for that sentence to be true or false, but which do not attempt to explain why certain patterns of inference are correct and others are not.

(2) What Dummett calls properly semantic theories of meaning, which do attempt to explain why some patterns of inference are right and others are not. In the terms of this essay, we can identify this with

the idea of a model-theoretic characterization of the truth conditions of propositions. Here the truth conditions of the propositions are explained as resulting from the semantic values of their components.

The key point here, in understanding Dummett's contrast, is that when we give a truth-theoretic characterization of the meanings of the sentences of a language, in Davidson's style or something like it, then logic of the object language is simply the logic of the metalanguage. The patterns of inference that are correct for the object language are simply those that are valid for the metalanguage. So the truth-theoretic characterization of the meanings of the object-language propositions cannot explain why certain patterns of inference are correct and others are not.

For a simple example of the distinction at work, consider the contrast between truth-theoretic characterizations of meanings of the propositional constants and characterizations of the meanings of the constants in terms of truth tables. At first this might seem like a subtle distinction. Is there any difference at all between the two? The truth-theoretic characterizations will contain axioms like:

'Ø & Ψ' is true if and only if Ø is true and Ψ is true.

What does this tell us about the inferential behavior of '&' in the object language? As the axiom is usually understood, it does nothing to explain the inferential behavior of '&'. All we can say is that '&' will behave in exactly the same way as does 'and', considered as a sign of the metalanguage. The inferential behavior assigned to '&' is simply a projection onto the object language of the inferential behavior of 'and' in the metalanguage.

Consider, in contrast, explaining the meaning of '&' by means of a truth table. Here we characterize the meaning of '&' in terms explicitly of a function from truth-values to truth-values. A set of introduction rules for '&' will then be justified by showing that it is the weakest possible set of introduction rules that from true inputs are guaranteed to yield a true output. A set of elimination rules for '&' will then be justified by showing that it is the strongest set of elimination rules still guaranteed to yield only truths as outputs from true inputs. This is not a matter of projecting the rules of inference for some sign in the metalanguage onto the object language. In fact there may be no sign corresponding to '&' in the metalanguage. Indeed the logic of the metalanguage may be quite different to the logic of the object language. So here we do not have an a priori guarantee that the inferential behavior of '&' will merely reflect the inferential behavior of some sign in the metalanguage. It is in that sense that the 'properly semantic' account of the meaning of the constant can be said to explain the rules of inference for it.

Another example is provided by modal statements. You might state the truth condition of 'Necessarily, Ø' by saying:

'Necessarily, Ø' is true if and only if necessarily, Ø.

But once again, this characterization does not explain why 'necessarily' exhibits the inferential behavior that it does. For example, we get no explanation of why it is that 'Possibly, Ø' follows from 'Necessarily, Ø'. All that happens is that the inferential behavior of the metalanguage term 'necessarily' is mirrored in the object language. To find an explanation of the correctness of the inferences here, we need to move to a model-theoretic approach that assigns a semantic value to Ø, such as a set of possible worlds at which it is true, and explains that 'Necessarily, Ø' is true if and only if Ø is true at all possible worlds. This way of explaining how the truth-values of the sentences in the object language are determined will give us an explanation of the validity of inferences involving those sentences. It will not merely be a matter of projecting onto the object language an exact mirroring of the inferential behavior of some term in the metalanguage. Indeed, in giving this kind of model theory, we need not even be using a metalanguage that has any of the ordinary modal terms in it.

On this way of understanding the truth-conditional approach to characterizing meaning, then, a grasp of truth conditions should provide you with some grasp of why correct patterns of inference are correct. An understanding of one's language should provide one, not only with a grasp of which patterns of inference are legitimate, but an understanding of why they are legitimate. This is where we most sharply see the difficulty in reconciling a truth-conditional account of meaning with functionalism. On the functionalist's account, all that the grasp of a term comes to is a capacity to operate with it in the right input–output transitions. These will include inferences involving the term, as well as, for example, transitions from perception to judgment and transitions from judgment to action. The trouble now is that a mere capacity for correct inference involving a term – a capacity to engage in correct transitions involving the term – however extended, cannot of itself constitute an understanding of *why* those patterns of inference are correct. Functionalism stops short, with a mere characterization of the transitions from content to content one does engage in.

You might point out that there is no question of a subject with *no* capacity to reason being provided with a grasp of some intended model for some central area of discourse, and then deriving from that a capacity for successful inference. The capacity to *derive* the correctness of rules of inference already presupposes some capacity to engage in correct reasoning. The role I am envisaging here for grasp of the intended model is not, however, that it should enable someone with no capacity at all for reasoning or any other transition in thought to become capable of it. We start in the middle: we begin with someone who already meets all the functional conditions for grasp of the terms, is capable of all the right input–output transitions, and we suppose them now to grasp the intended model for this type of discourse. You might object that this procedure will be circular, and that it can validate only the rules of inference that the subject is already using. But that wouldn't be right, as we have seen already. That was the whole point of working through the sense in which a properly semantic theory for an area of discourse is not bound simply to validate whatever rules of inference are already being used in the metalanguage. In the various cases we discussed, we saw that there is no guarantee that the method by which the semantic theory justifies the patterns of transition for this area of discourse will validate just the patterns of inference that the subject is already using.

Of course, when one thinks of the subject using reasoning in some metalanguage to validate patterns of reasoning in an object language, we are thinking of someone operating at a level of sophistication that may not be credibly ascribed to just anyone who understands a language. But we can, I think, envisage a much lower-level way in which grasp of the intended model could be justifying one's use of a pattern of reasoning. The more general, schematic way of describing the way in which the subject's use of particular patterns of reasoning is justified by grasp of the intended model is to say that we are dealing with a subject whose use of particular patterns of reasoning is *causally sustained* by their grasp of the intended model. One form this causal sustaining can take is when the subject explicitly reflects on the rules of inference they are using and how they are validated by the intended model. However, there is no reason in principle why a subject should not exhibit a formally similar causal sensitivity even in the absence of explicit reasoning on the subject's part. All we need is to be able to say that the subject's employment of particular patterns of use of the terms is causally sensitive, in the right kind of way, to the subject's grasp of the intended model.

It used to be acknowledged that 'narrow' functional role – functional role limited to inputs and outputs local to the subject's body – did not provide an account of truth-conditional content, but that 'broad', or 'long-arm' functional role, which included causal inputs from distal stimuli in the subject's environment, and the subject's causal impact on distal stimuli, would provide an account of truth-conditional content. If the line of argument I am setting out is correct, though, then this plainly cannot help. An account of the subject's grasp of truth-conditional content should, I am suggesting, say something about the subject's understanding of why the patterns of transition from state to state that the subject goes in for are correct. An appeal to long-arm functional role adds nothing to the point: it adds nothing but more input–output transitions, beyond the confines of the subject's body to be sure, but simply more input–output transitions, while doing nothing to explain why any of those input–output transitions, confined to the subject's body or not, are correct.

This point is often obscured in expositions of the idea of long-arm functional role by the remark that the distal object mentioned in specifying the long-arm functional role is the 'referent' of the term. This can be confusing, because it can obscure the fact that all that is going on is that more functional role is being specified. There is nothing different about the way the subject meets this aspect of the functional-role specification that could show how the subject grasps why the patterns of functional role being used are correct. If we are to show how a grasp of the referents of the signs being used could constitute the subject's grasp of what justifies the use of certain patterns of inference, then we need to say something about the subject's cognitive grasp of reference, something that does more than merely giving us more functional role.

All these points apply to the case of perceptual demonstratives just as much as they do to any other sector of language. If a grasp of truth conditions for demonstrative discourse is to constitute grasp of that which justifies the employment of particular patterns of use in connection with demonstratives, then this grasp of truth conditions cannot be explained as a matter merely of conforming to a particular functional characterization, not even a long-arm functional characterization. We have to explain what it is that could contribute to one's understanding of perceptual demonstratives in such a way that it could be said to confront one with the intended model of the discourse, the prototype of all admissible models for the discourse, in such a way as to provide one with some grasp of the justification for the employment of a particular pattern of use for the demonstratives. The natural answer is that in perception we are confronted with the references of perceptual-demonstrative terms, and to that extent we can be said to perceive the intended model for demonstrative discourse. But if we think of perception as a matter merely of causal connections between objects in the environment and the sensory organs of the subject, then perception can add nothing but more functional role.

To characterize the sense in which we have knowledge of the references of demonstrative terms that can provide us with a properly semantic grasp of demonstrative discourse, we have to appeal to the idea that we *experience* the references of demonstrative terms. In ordinary perception we are not just causally affected by the references of demonstrative terms. We have sensory awareness of the objects being referred to, and it is our sensory awareness of the objects referred to that provides us with our grasp of the intended model for perceptual-demonstrative discourse. It is our sensory awareness of the objects referred to that causally sustains our employment of particular patterns of use of demonstrative terms.

It is often said that there are two fundamental problems facing functionalism as a theory of mental states. One is that functionalism can't give an account of the sense in which our grasp of meaning is *normative*, that it is not just that we do in fact conform to certain patterns in our thinking and talking, but that we have some grasp of why it is *right* to think and reason as we do. The other fundamental problem is *sensory awareness* (the problem of 'qualia'). The idea is that states of sensory awareness have, as it were, intrinsic phenomenal content that can't be analyzed in terms merely of input–output patterns in the subject's mental life. I am proposing that these two limitations of functionalism are connected, that the fundamental reason why functionalism can't give an account of the normative dimension of perceptual-demonstrative discourse in particular is that it can't give an account of our sensory experience of the objects and properties around us.

3 DUMMETT'S OBJECTIONS TO THE IDEA OF 'GRASP OF THE INTENDED MODEL'

The reason I began with what Dummett calls the common explanation of the Gödel theorem is not that I want to defend that view of our understanding of arithmetic. The reason I set it out is, rather, that an account like this seems quite compelling for our understanding of statements about the ordinary concrete world that we see and hear, the world of tables and chairs, people and trees, seas and mountains, and so on. In this case we do not have to talk about a quasi-perceptual understanding of the intended model of our discourse. You do literally perceive the intended model. The objects and properties that your discourse concerns are staring you in the face. So the picture of understanding that seems natural for arithmetic seems not only natural, but compelling for our understanding of statements about the medium-sized world in which we ordinarily live.

Having articulated the 'common explanation' of the Gödel phenomenon, Dummett raises two objections to the naive picture. The first is that it makes our understanding of mathematics incommunicable; you can never know how someone else is interpreting a mathematical statement, since you cannot tell which mathematical domain the other person is surveying, or quasi-surveying, to provide them with knowledge of the right model.

A natural objection is that, since I cannot look into another man's mind in order to read there what meaning he attaches to 'natural number', since all I have to go on is the use which he makes of this expression, I can never know for certain that he attaches the same meaning to it as I do.

Does this objection apply to the idea that we understand propositions about the concrete world by perceiving the intended model? The question puts some pressure on our analysis of perception. If, for example, you think of perception as a matter of having sensations, then you might think there is a problem about whether we can know that we are having the same sensations. But on the face of it, we can know that we are perceiving the same domain. The shared world that you and I jointly perceive is just the world in front of us, and we can check, in all the usual ways, whether we are looking at the same sector of reality. ('Do you mean that man at the news-stand?' 'Do you mean my right or your right?', and so on.) So the problem about incommunicability does not, on the face of it, arise in this case. I shall discuss this point a bit further after looking at Dummett's second, and perhaps more fundamental, objection.

Dummett's second objection to the common explanation of the Gödel phenomenon is that it appeals to a quasi-perceptual 'intuitive grasp' of the intended model without explaining what it is to access the model in this way. After all, what can it mean to be 'given' the model, except that we have a description of it? But such a description is exactly what the axioms of the formal theory provide. If there is something more to 'access' than merely being given the axioms of the theory and the rules of inference that tell one how to derive the implications of those axioms, we have not yet been told what that is. But there must be more to it than this, if the common explanation is to explain how one can recognize the truth of the Gödel sentence.

The account of Gödel's theorem we are considering, however, operates with the notion of a model as if it were something that could be given to us independently of any description: as a kind of intuitive conception which we can survey in its entirety in our mind's eye, even though we can find no description which determines it uniquely. This has nothing to do with the concept of a model as that concept is legitimately used in mathematics. There is no way in which we can be 'given' a model save by being given a description of that model. If we cannot be given a complete characterization of a model for number theory, then there is no way in which, in the absence of such a complete description, we could nevertheless somehow gain a complete conception of its structure.

A parallel question can be raised for the perceptual analysis of our understanding of ordinary statements about concrete objects. In what sense is the world 'given' to us in perception? Doesn't perception merely provide us with further descriptions of the world? The question this raises for the analysis of perception is whether we have to think of perceptual experience as a matter of merely being given yet another description of our surroundings. There is something of a crux at this point. On the one hand, you might indeed think of perception in *representational* terms, as a matter of having representations of your surroundings. In that case it will indeed seem that perception offers nothing but more representations, and that consciousness cannot provide a way of 'grasping the intended model' of demonstrative discourse that could provide one with a grasp of why the pattern of one's use of demonstratives is correct. Of course, the representations one has in conscious experience will be subject to their own patterns of use, including patterns of use for demonstratives. But that fact could not of itself supply the reason why any one pattern of use is correct. On the other hand, you might think of perceptual experience as a matter of having, perhaps in addition to various perceptual representations, *sensations* of the world around you. But this model of perceptual experience as a matter of having sensations simply articulates a conception of experience on which it provides no epistemic access to the world around you, a view of perceptual experience as providing knowledge only of itself. There is, however, an alternative to thinking of perceptual experience as representation or of thinking of it as sensation. We can think of perception as fundamentally a *relation* to the external scene, and that will give a picture of perceptual experience as providing just the kind of access to the intended model that can provide our understanding of language.

If we think of perceptual experience as a relation to the external scene, we still have to observe Dummett's point that there is such a thing as the *way* in which the external scene is given to one in perception. We cannot say simply that the perceptual experience consists in one's being related to various objects and properties around one, without saying *how* the experience relates one to those objects and properties. Of course, if you think that the notion of a *way* of being related to an object or property can be explained only by appealing to the idea of a descriptive identification of the object or property, or a particular kind of sensation induced by the object or property, then we will have made little progress beyond the models of perception as representation or as sensation. But I want to propose that we can do better than that. We can explain the experiential relation in terms of its attentional structure; and we can characterize attentional structure in terms of the way in which the objects and properties in the scene contribute to the causal structure of the perceiver's experience.

In an interesting series of papers, Sebastian Watzl has stressed the centrality of attentional structure to the phenomenology of experience (see e.g. Watzl, forthcoming). Watzl suggests in effect that the way to characterize the phenomenology of attention may be to introduce some primitive notions, such as 'x is more peripheral to attention than y', that we can use to explain ideas such as 'y is the focus of attention'. I think that if one thinks of how one would naturally give first-person reports of one's experience that reflect the role of attention, that is how one would naturally proceed. But what I will suggest here is that there is another approach we can take to characterizing the phenomenology of attention. We can think of describing the phenomenology of attention as a matter of describing the causal structure of one's experience of the objects and properties around one. That is, describing the phenomenology of attention is a matter of describing how one's experiences of those objects and properties are causally related to one another in different kinds of ways.

We can begin on this by remarking a distinction between two ways in which a perceived property can function in relation to an object or region. First, the property may be used to *select* the object or region. Secondly, the property may be *accessed* as a property of that object or region. Selection is what makes the object or region visible in the first place; selection is what makes it possible for the subject to focus on that object or region in order to ascertain its various properties. Access is a matter of the subject making it explicit, in one way or another, just which manifold properties the object or region has.

This distinction is written large in Huang and Pashler's (2007) analysis of visual attention, which emphasizes a distinction between *selection* and *access*. (In *Reference and Consciousness*, I explained this kind of distinction under the heading, 'The Double Use of Feature Maps'; see Campbell

2002.) They presuppose an underlying architecture of Treismanian feature maps, in which the locations of various features of different sorts are plotted: the locations of various colors, shapes, orientations, movements, and so on. In *selection*, a particular property – say, a particular color – is used to identify a particular region or object. So, for example, an area might be selected as the locus of all the redness in the observed scene. Huang and Pashler say that what is being selected is: 'a collection of locations. It should be emphasized that those locations are regions exactly covered by relevant stimuli (i.e., one should imagine the map as being, as it were, shrink-wrapped to conform tightly to the object)' (2007, 601n2). I do not think, though, that we should see them as taking an undefended stand on the idea that attention is always to location rather than to objects. Rather, we could understand them as saying merely that when an object is selected, it is selected by selecting the region it currently occupies.

In *access*, the various maps are interrogated again, now to make explicit the various properties of this selected region or object. Huang and Pashler diagram the situation as shown in Figure 5.1.

As you can see, Huang and Pashler have a proposal as to where perceptual experience fits into this picture: they suggest that it is a consequence of the kind of processing they describe that the accessed properties of a selected region will be contents of consciousness, aspects of the phenomenal experience of the subject. Now it seems to me that the empirical basis for Huang and Pashler's distinction between selection and access, and their way of locating the distinction against a broadly Treismanian background, is forcefully and persuasively made out. But their way of locating sensory awareness in the picture seems to me questionable. It seems arguable that properties that are used in selection, but not accessed by the subject, may themselves figure as contents of experience.

There are many cases in which a property is used as the basis for selection of an object or region and may not be accessed, or even accessible, by the subject. In such cases, there may be a compelling case to be made that the property is entering into the phenomenal content of the subject's experience. If that's right, then it will be a mistake to tie awareness to access; the way to connect awareness to attention is, rather, to connect awareness of a property to the possibility of that property's being used as a basis for selection of an object or region.

Suppose we consider a particularly simple kind of example: the use of tests for color vision in which all that separates a particular figure from its background is the hue. So, for example, you may be presented with an array of variously colored blobs of varying luminance in which all that

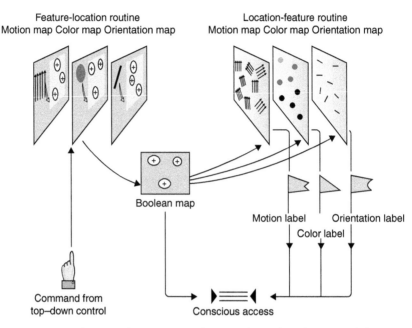

Figure 5.1 Boolean map theory. Note: Boolean map theory depends on an underlying architecture of Treismanian feature maps. In the example shown, a feature, redness, is used to select all and only the red areas. A first consultation of the feature map is made, to find which regions have red at them. This yields a Boolean map, which at this stage is not thought of as labeled. The underlying feature maps are then consulted again to determine what features are found at the region selected, and the labeling then makes those features explicit.
Source: Huang and Pashler 2007.

there is to systematically separate a figure 5 from its background are the colors of the various blobs. In such a case, it is hard to imagine how you could see the 5 without having conscious experience of the various colors involved. If you did not have conscious experience of the various colors, you might perhaps have a hunch that there is a 5 there in the scene. You might, if asked to engage in blindsight-style forced-choice guessing as to which figure is present, actually guess that the number in the scene is a 5. But if the 5 is visibly there, as a 5 can be present in ordinary vision, then we cannot imagine how that could be unless you had phenomenal awareness of color. It is not exactly that there is a contradiction in the idea of seeing the 5 in this kind of case, without experiencing its color. The problem is rather that the experience seems entirely unimaginable; we can make nothing of the idea of such an experience.

Still, although you may have used your conscious experience of the color to select the 5, and can then go on to access various properties of the 5, such as its size, shape, and orientation, it seems entirely possible that you could select the 5 on the basis of color without yet having any capacity to access color properties. This is an entirely logical point about the mechanics of selection and access. Merely having the capacity to select objects of regions on the basis of color does not of itself automatically imply that one must have a capacity to access color properties. And it is not difficult to find illustrations of the point. Human children have color vision in place long before they have anything in the way of a color vocabulary; it is entirely possible that a child a few months old could see the 5 in the kind of display I am describing, without having any ability to give a verbal report of its color. You might object that verbal report is only one way in which access to a seen property could manifest itself. That is correct; there are other ways in which access to color could show up. For example, it could be that one is capable of color induction: having been exposed to red berries that were not good to eat, one is in the future reluctant to eat red berries. But the same point applies here. A mere ability to see the 5 against its background does not yet imply that one has a capacity for color induction. Nor, to take some further obvious examples, does it imply that one has a capacity for color matching, or a capacity for ordering objects by their colors. If one sees the 5 then, I have said, one must have experience of its color; but that experience of color may be quite deeply recessive, and not show up in any way other than in a capacity to use it to distinguish objects from their backgrounds.

I think we can use this kind of point to begin articulating what we might call the 'attentional structure' of someone's experience of a scene. Experience of the object – the 5 – is made possible by, causally facilitated by, experience of the color of the object. If you can report on the shape of the object – if you can say that it is a 5, for example – then your experience of the object (caused by your experience of the color) and your experience of the shape of the object are jointly playing a role in causing your verbal report. But there is a distinction between the causal roles of your experience of the color and your experience of the shape of the object, in this case. Your experience of the color is not (we may suppose, in your case) capable of causing a verbal report of the color of the object, even though you are experiencing the 5. In contrast, your experience of the shape of the object is capable of causing a verbal report of the shape of the object in the context of your experience of the 5. But your experience of the shape of the object is not what is causally facilitating your experience of the object itself; it is the color that is doing that work.

Similarly, there is an asymmetric causal dependence between your experience of the object and your experience of the shape of the object, in the roles they play in verbal report. It is only because you are experiencing the object that you can experience its shape in such a way as to cause verbal report of its shape. You could, however, experience the object without experiencing its shape, in such a way as to cause you to have the capacity to make verbal reports of other aspects of the object. But there is no such thing as holding on to consciousness of the shape in the absence of experience of the object, in such a way as to allow you to make verbal reports as to the shapes of other objects. I am dealing here with a deliberately simple example; obviously, articulating the causal-attentional structure of one's experience of a complex scene over a period of time would be a more difficult business. But it seems in principle possible to do this kind of analysis for one's experience of any scene, however complex and over however long a period.

On this approach, I think we can see how to address Dummett's concern about the idea of being 'given' a model otherwise than by way of a description of it. I don't here want to dispute Dummett's point for the arithmetical case, which is the only case he was concerned with; but I do want to suggest that we can address his point for the case of discourse involving perceptual demonstratives. To characterize someone's experience of a scene, I am suggesting, it is not enough merely that we characterize the objects and properties; we have to characterize the causal roles that the various objects and properties are playing in the subject's experience of the scene. This is not a matter of saying something about how the subject is *representing*, or describing, the scene. Nor is it a matter of saying something about the sensations the subject is experiencing on looking at the scene. Talk of the subject's 'representation' or 'description' of the scene would suggest that we describe things in terms of the way the subject is accessing the properties of the objects observed; and it should be amply clear that the kind of articulation of causal structure that I am describing is operating at a deeper level than that. But we can talk about the 'way' in which the various objects and properties in the scene are 'given' to the subject. For example, in the case I described of seeing the figure 5 in an array of blobs, we can naturally think of the 'way' in which the figure 5 is given to the subject as characterized by the color of the blobs differentiating the figure from its background. But this is not a matter of the subject grasping some representation of color, because as I said, the subject may have no capacity to access the color of the 5 at all.

We can, then, explain a notion of 'mode of presentation' of an object as follows: the visual mode of presentation of an object is the properties of the object that make it causally possible for you to perceive the object. In our simple example of the figure 5, the color and location of the 5 will play a role in making it possible for you to see the figure 5. The shape of the 5 is not playing such a role in making it possible for you to see the 5. So although both color and shape are visible properties of the 5, color is part of the mode of presentation of the object and shape is not.

This lets us see how to answer our question about when trading on identity is legitimate, in terms of the causal structure of the subject's experience of the object. Consider an inference such as 'This 5 is *F*; this 5 is *G*; hence, something is both *F* and *G*.' We can imagine a case in which this inference is not valid, even though it is the same figure that is being referred to both times; suppose, for example, that we are dealing with a solid wooden figure 5, and that the first premise involves seeing that figure 5 on the basis of its color, whereas the second premise involves manually grasping the figure 5, singling it out by its weight and shape. In that case the inference is not valid as it stands; there are admissible models in which the two demonstratives refer to different things. Suppose, however, that we are dealing with a case in which the demonstratives in the two premises are understood in the same way: that is, in both cases it is the same properties, let us say color, that are used to select the 5 as figure against ground. In that case, since your grasp of the intended model involves understanding the demonstratives in the same ways, any admissible model of the premises must assign the same references to the two terms, and the inference is valid: in any admissible model in which the premises are true, the conclusion is true.

4 KEEPING TRACK

So far I have discussed perceptual demonstratives without giving any consideration to the dynamics of object perception: to our capacity to keep track of objects across variation in their perceptible characteristics, including their location with respect to the perceiver. This is such a pervasive characteristic of object perception, however, that we can hardly conclude without reflecting on its impact on demonstrative reference and the validity of demonstrative inference. Suppose, for example, that you are watching an object move around a room. You may be using the color and location of the thing to select it from its background. Nonetheless, the color and location of the thing may be changing continuously over time.

And it may be manifestly the same thing over that period of time. The sameness of the object may be just as manifest as it would have been had there been no change of color and no change of location. If trading on identity is legitimate in the case in which the same color property is used twice to select the object, then trading on identity should be legitimate in the case in which the object changes color and location while remaining manifestly the same.

The trouble is to explain the sense in which the object is 'manifestly the same' despite the change of color and location. After all, there certainly are cases in which a change of color and location would mean that you lose the right to take it to be the same object. Suppose for example that you are blindfolded and that while you are blindfolded the thing changes color and moves to a new location. When the blindfold is removed there it is, with its new color at the new place. Is it the same thing? Here you really may need to appeal to auxiliary premises and reasoning to establish the identity of the object, rather than simply trading on identity. Similarly, there are subtler cases in which no blindfold is involved, but you simply do not keep track of the object. Your vigilance flags and now there is an object of a different color in a new place. Is it the same one? Here again you may have to appeal to auxiliary premises and reasoning to establish the identity of the object, rather than merely trading on identity. So what exactly are we doing in keeping track? There are two cases. In one, you keep track of the object through the change in color and location, and trading on identity is legitimate. In the other, you do not keep track of the object, but you do see the old color and location, and you do see the new color and location. Here, trading on identity is not legitimate. So what is going on when you do 'keep track'? Is there any more we can say, other than that in cases where you do keep track the trading on identity is legitimate?

One aspect of the question here has to do with the *mechanism* by which we do keep track of objects. But before going on to that, it is worth articulating a little more just what the phenomenon is. Suppose we consider a period of time over which you are keeping track of an object. At the beginning, at time t_1, you can use the color of the object to select it; having done so, you can access its further properties, such as its shape and orientation, for example. So let's suppose that you form the judgment, 'It's a figure 5.' At the end of the period we are considering, at time t_2, during which the color of the object has changed, you use the new color of the object to select it from its background, and form the judgment, 'It's tilted to the right.' Since you have kept track of the object over that period, you can immediately trade on the identity, and draw the conclusion, 'There's a figure 5, tilted to the right.' If we kept strictly with the account of modes of presentation that we had reached at the end of the last section, we should have to say that there were different modes of presentation here, since the object was being selected in different ways at time t_1 and at time t_2. While that seems a correct account of the case in which you did not keep track, it does not seem like an adequate description of the case in which you did keep track; something has been missed out, something that is giving you the right to trade on identity.

Over many years, Zenon Pylyshyn has studied multiple-object track-ing: the ability of perceivers to keep track of various kinds of moving objects. In a classical multiple-object-tracking task, you are presented with an array of visual targets, four of which flash briefly, to indicate that they are 'yours'. Then all the targets start moving. At the end of the trial, one target flashes and you are asked whether it was one of yours. People are well above chance in doing this, though not perfect. Pylyshyn argues that the way in which we manage to keep track of objects is by assigning them 'visual indexes', of which the visual system has access to four or five. On the face of it, there is no reason why the analysis pro-vided by Huang and Pashler could not allow for the existence of visual indexes, which would be used to keep track of an object once selected, and its properties accessed, until the object is reselected and its proper-ties accessed again.

Of course, if we put the idea of visual indexes like that, there is a ques-tion as to how much empirical content the theory has. What, if anything, does the hypothesis of visual indexes add to the observation that we can keep track of objects over time? In fact, as Pylyshyn expounds it, the idea would add a lot. But before going into the details of Pylyshyn's view, sup-pose that, just for the moment, we regard the notion of a 'visual index' as a placeholder for an empirical account of how we keep track of objects over time, with details of how its biological implementation and role in information processing to be filled in later.

Suppose that, over a period of time, from t_1 to t_2, you are keeping track of an object by means of a visual index. On the face of it, there are three separate tasks here:

(1) Selecting the object at t_1, so that you can access its various features;
(2) Keeping track of the object from t_1 to t_2, so that it is manifestly the same; and
(3) Selecting the object at t_2, so that you can access its various features.

These all seem to be different tasks. On the face of it, selection for access in task (1) could be accomplished just as Huang and Pashler suppose, using a feature such as color to select the object by selecting the region it occupies. Then there could be a visual index, operating in something like the way Pylyshyn supposes, between time t_1 and time t_2. Then selection of the object at time t_2 could again proceed in the way Huang and Pashler suppose, so that again the object is selected by selecting the region it occupies. On this picture, we can generalize the account of trading on identity that I gave in the last section. We can say that trading on identity between two uses of a demonstrative is legitimate when either:

(a) it is the same properties that are used to select the object as figure from ground, or
(b) the uses of the demonstrative reflect selections of an object as figure from ground that are linked by the successful use of a visual index to keep track of the thing.

Notice that it has to be 'successful' use of the visual index since legitimate trading on identity requires that the object actually be the same, not just that the subject have the right to take it to be the same.

There is a sharp contrast between this picture and Pylyshyn's, however. In effect, Pylyshyn proposes that the visual index can do the work described under (1) and (3) above, as well as the work described under (2) above. This means that on Pylyshyn's picture, there is no evident role in visual reference for an underlying array of Treismanian feature maps, and no place for the idea that an object is selected for access by selecting the region it currently occupies. Rather, the selection of an object for access is achieved merely by assigning it a visual index. Once the visual index has been assigned, that alone means that it is possible to access the various properties of the object. And locations, or the region occupied by the object, do not play any particular role in this process; location does not play the central role that Huang and Pashler give it.

This is a radical alternative to the picture I have set out so far; the suggestion would be that we do not use the notion of 'the property experience of which allows you to select the object as figure from ground', in giving an account of trading on identity; we just appeal to the idea of a visual index being assigned to an object. On Pylyshyn's account, there would still have to be some account given of the kinds of properties of an object that allow it to be assigned a visual index in the first place, and to sustain tracking by means of a visual index (for a review of some of the work on this topic see e.g. Pylyshyn 2007; Scholl 2007). But these properties actually play no role in explaining what makes trading on identity legitimate; all that matters for that question is that the index has been assigned to the object and that it is the same index over a period.

The issues here are empirical, and I think it is fair to say that the radical version of Pylyshyn's theory, which is what he himself endorses, is not well supported. That is, there is little evidence that assigning a visual index itself is enough to allow interrogation of the stimulus to access its various properties; the evidence is that we need something like the focus on location implied by Huang and Pashler's account, with its background, to appeal to an underlying level of Treismanian feature maps, to explain how it is that one can visually access the various properties of a single object (tasks (1) and (3) above). There are a number of telling findings here that are due to Pylyshyn himself. Suppose, for example, that at the start of the multiple-object-tracking trial, your four targets are all labeled with different numbers; then at the end, when one flashes, you are asked, not merely whether it was one of yours, but which number it had. This is a much harder task than multiple-object tracking itself, and people generally cannot do it. What this seems to show is that visual indexes function merely to achieve task (2) above: to keep track of objects. They do not of themselves allow one to interrogate the stimulus for its properties, and to keep track of it as the thing that has certain properties. Similarly, Pylyshyn himself notes that changes in the properties of the object as one is keeping track of it, not only do not interfere with the capacity to keep track of it, but may simply not be noticed by the subject who is nonetheless successfully keeping track of the thing. This again makes the point that allocation of the visual index is one thing, and the capacity to access multiple properties of the object is another.

So it seems to me that for the moment, at any rate, we should work with the assumption that selection for access is one thing, and the assignment of visual indexes for tracking is another. That means we keep, as our canonical statement of when trading on identity is legitimate, that it is legitimate when (a) it is the same properties that are used to select the object as figure from ground, or (b) the uses of the demonstrative reflect selections of an object as figure from ground that are linked by the successful use of a visual index to keep track of the thing.

5 THE EPISTEMIC ROLE OF CONSCIOUSNESS

If the line of thought I have been pursuing in this chapter is correct, then there is an epistemic role for consciousness, for sensory awareness in particular, in our grasp of meaning. It is sensory awareness, I have been suggesting, that provides us with our grasp of the intended model for perceptual-demonstrative discourse. And since our grasp of the intended model provides us with our prototype of an admissible model, it is our grasp of the intended model that provides us with our understanding of why some inferences are valid and others are not. So sensory awareness, by providing us with our grasp of the intended model for our discourse, provides us with our insight into why some inferences are correct and others are not. As I remarked earlier, Dummett objects to this kind of account in the mathematical case, arguing that it would mean that we can never know just how another person is understanding arithmetic, since we can never 'look into the other person's mind' to find what model they are associating with the language. The point of contrast with the perceptual case, however, is that here we do not need to 'look into the other person's mind' to see which sector of reality they are associating with the terms they use. The direction of dependence is round the other way. It is because you are able to observe the same environment as the other person that you can know what is in their mind. When you are jointly attending, with someone else, to some aspect or aspects of your surroundings, the first phase is one in which you are aware merely of *which* things your companion is attending to. *How* the other person is attending to them, what kind of perspective the other person has on their surroundings, comes later. This seems to be so, both logically and developmentally (cf. Moll and Meltzoff, forthcoming).

I think it is arguable that this epistemic role for consciousness suggests something that is distinctive of conceptual thought of the sort typically employed by humans. Consider, in contrast, the honeybee. A lot is known about what in some broad sense we might call the 'cognition' of the honeybee. We know that honeybees represent the times at which things happen – von Frisch was famously struck by the regular appearance of honeybees on his balcony at breakfast time – and can track the possibilities and limitation of the ways in which they represent time. We know that honeybees can represent the locations of pollen sources, and that they can communicate to one another where the pollen is to be found. But are honeybees conscious? That is really a moot point. I used to talk regularly with a colleague in zoology, and sometimes I would explain to him various hypotheses about the representational or functional architectures that philosophers supposed to be constitutive of consciousness. 'Well if that were right', he would say with some finality, 'then honeybees would be conscious'. Since he worked on honeybees, I tended to give way, rather uneasily, to his authority as to whether they were conscious. For the moment, I don't want to pursue the question whether honeybees are conscious. The point that matters is that we pursue the study of honeybee cognition quite independently of any assumptions as to whether they are conscious. In fact, suppose that honeybees are conscious. Suppose that they do experience, for example, pangs and pains and delight. Their experiences may actually not matter very much for the study of honeybee cognition. They may, for example, fall far short, in richness and complexity, of what is there in honeybee cognition.

Suppose there are Martian biologists, and that they are studying human beings. Martians are, as is well known, much larger than human beings; in terms of physical size, they stand to us rather as we stand to honeybees. They are also phenomenally clever, their intelligence stands to ours somewhat as our intelligence stands to that of the honeybee. The Martians study human cognition. They are struck by the complexity of, for example, our mental representations of time. Any human can hold the outline temporal detail of a life, as well as endless schemata relating the order of repeated activities, and a capacity to feel the pulse of music, to recognize humor in a rhythm, and so on. Humans are also capable of many layers of spatial representation. Anyone navigating a familiar town, for example, may be using mental maps of where everything is in many different reference frames and at many different scales. Martians have achieved considerable insight into the scope and limitations of our ways of representing time and space. As to whether humans are conscious, they regard that as rather a moot point. Some of them attempt to give analyses of consciousness as a matter of possessing certain representational capacities, or functional architectures; their colleagues respond cuttingly by saying, 'Well, if that were so, then humans would be conscious.' However, in the end they do recognize that it is hard to find a compelling answer one way or the other as to whether humans are conscious. Their take on human cognition is very much like our take on honeybee cognition. They pursue its study with a great deal of energy, ingenuity, and discipline. They have achieved a considerable understanding of much of what goes on in human cognition, and they have related it to the functioning of the brain. This achievement is independent of them having even raised the question whether humans are conscious. Moreover, suppose humans do have consciousness. This fact might not matter very much for the cognitive lives of humans, according to the Martians. In the case of the honeybees, the throbs and pulses of pleasure and pain may have nothing of the richness and complexity of bee cognition. So too in the case of the humans, the Martians argue, the throbs and pulses of human consciousness, if there is such a thing, may have nothing much to do with the marvelous richness and structure of human cognition.

There is an austere and illiberal reading of recent cognitive science that remarks on its traditional exclusion of notions of consciousness from serious scientific work, that in effect proposes an analysis of human cognition that is very much like that proposed by our Martian scientists. On this reading, the working assumption is that the notion of consciousness is not of much use in understanding human cognition. The assumption is not exactly that no such thing exists. The point is rather that in understanding and explaining human cognition, the phenomenon of consciousness does not matter much. The nature of human consciousness may have nothing of the richness and complexity of ordinary human cognition, so far as the methods of classical cognitive science are concerned. Even though humans are conscious, the fact has little bearing on their knowledge of their surroundings.

This austere and illiberal reading of classical cognitive science contrasts sharply with the methodology of traditional epistemology. Traditional epistemology took it that sensory experience was the basis of our ordinary knowledge of our surroundings. In fact, it is today still often taken to be a commonplace truth that it is only because we have sensory experience that we have knowledge of our surroundings. This shows up in the descriptions that we give of phenomena that involve perception without awareness. We usually take it that perception in the absence of sensory experience cannot yield non-inferential knowledge of your surroundings. Consider the phenomenon of 'blindsight'. Blindsight patients have damage to their visual system as a result of which they have visual experience in only one half of the visual field. They have no awareness of stimuli presented to the blind field. But, when asked to guess about the characteristics of a stimulus in the blind field, they are, for many observable characteristics, reliably correct. Despite their reliability, these subjects are always said, by themselves and by the researchers testing them, to be 'guessing'. We do not in practice take them to have non-inferential knowledge of what is in the blind field. When there is conscious awareness of the stimulus, however, as in ordinary vision, we take it as unproblematic that the subject has non-inferential knowledge of what they see.

We do in practice take it that experience is required for knowledge.

This dependence of knowledge on experience does not apply only to propositional knowledge. Consider knowledge of which things and properties are to be found in your surroundings. For example, consider knowledge of the colors. I don't just mean knowledge that there is such a phenomenon as color, but knowledge of which property scarlet is, for example. We usually find it compelling that this knowledge can be provided only by experience of the colors. A capacity for correct, blind-sight-style guessing as to the colors of things would not be enough for knowledge of what the colors are; you wouldn't know what scarlet is, in the absence of experience of it. Or again, we take it that to know which particular thing someone is trying to draw your attention to, you have to experience the thing; a mere capacity for blindsight-style guessing, a mere hunch that something is there, isn't enough.

When the honeybee processes the information it has about its surroundings, finding flight paths home or ways of communicating the information it has about sources of pollen, it is engaging in transitions that are validated, if at all, by their adaptive value. The honeybee itself has no grasp of the validation of those transitions. When it processes information, it does so blindly. You might argue that the reasoning we engage in as part of human conceptual thought is just like this. When we engage in particular patterns of reasoning, you might say, we do so blindly. We have no insight into why those patterns of reasoning are correct. If they were challenged, we should have nothing to say, except perhaps, 'Here we have a hitherto unknown form of madness.' But if the perspective I have been recommending in this chapter is correct, we do in fact have some insight into why the transitions we make in our reasoning about the perceived world are correct. Our sensory awareness of our surroundings provides us with knowledge of the intended model of our discourse, and thereby provides us with an insight into the validity of our inferences that is missing from honeybee cognition.

REFERENCES

Campbell, John (2002) *Reference and Consciousness*. Oxford University Press.
Dummett, Michael (1978) The philosophical significance of Gödel's theorem. In *Truth and Other Enigmas*. London: Gerald Duckworth & Co., pp. 186–201.
Huang, Liqiang and Pashler, Harold (2007) A Boolean map theory of visual attention. *Psychological Review* 114: 599–631.
Moll, Henrike and Meltzoff, Andrew (Forthcoming) Perspective-taking and its foundation in joint attention. In Naomi Eilan, Hemdat Lerman, and Johannes Roessler (eds.) *Perception, Causation, and Objectivity: Issues in Philosophy and Psychology*. Oxford University Press.
Pylyshyn, Zenon W. (2007) *Things and Places: How the Mind Connects with the World*. Cambridge, MA: MIT Press.

Scholl, Brian J. (2007) Object persistence in philosophy and psychology. *Mind &*
Language 22, no. 5: 563–591.

Watzl, Sebastian (Forthcoming) Attention as structuring the stream of consciousness. In Christopher Mole, Declan Smithies, and Wayne Wu (eds.) *Attention:*
Philosophical and Psychological Essays. New York: Oxford University Press,
pp. 145–173.

CHAPTER 6

Individuation, reference, and sortal terms

E. J. Lowe

In previous work (Lowe 2007), I have defended a thesis that I call *categorialism* regarding the individuation of objects, in the cognitive sense of the term 'individuation'. Individuation in this sense – which is to be distinguished from individuation in the metaphysical sense (Lowe 2003) – is the *singling out of an object in thought*. According to categorialism, a thinker can single out an object in this way only if he or she grasps, at least implicitly, some categorial concept under which he or she conceives the object in question to fall – such a concept being one that supplies a distinctive *criterion of identity* for objects conceived to fall under it. Plausible examples of such categorial concepts would be the concepts of an *animal*, a *material artefact*, and (what I shall call) a *geographical prominence*.

Categorialism, thus, is a more liberal doctrine than *sortalism* – the latter doctrine maintaining that an object can be singled out in thought only when conceived of as falling under some specific *sortal* concept, such as, for example, the concept of a *cat*, a *table*, or a *mountain*. As these everyday examples illustrate, categorial concepts are more abstract than any of the more specific sortal concepts that fall within their range of application: *animal*, for instance, is more abstract than either *cat* or *dog*, and *geographical prominence* is more abstract than either *mountain* or *island*. All sortal concepts falling within the range of application of the same categorial concept are, it seems clear, necessarily associated with the same criterion of identity, but they evidently differ with respect to the more specific *sortal persistence conditions* governing objects that fall under them. These sortal persistence conditions – which impose restrictions on what varieties of natural change an object can be supposed to survive while continuing to fall under the relevant sortal concept – are for the most part discoverable only empirically, whereas criteria of identity proper are most plausibly classified as relatively a priori metaphysical principles.

In the present chapter, I shall offer further arguments in support of categorialism and then go on to inquire whether these arguments can be extended from the domain of singular thought to that of singular linguistic reference: that is, I shall inquire whether it can reasonably be contended that a speaker cannot successfully refer to an object by means of a proper name unless he or she grasps, at least implicitly, that the name's referent falls under a certain categorial concept, which supplies a criterion of identity for the referent. This contention conflicts, of course, with the assumptions of any purely 'direct' theory of reference, to the extent that it makes an object's known satisfaction of some broadly descriptive specification a necessary – albeit not a sufficient – condition for successful linguistic reference to that object.

I SORTAL, CATEGORIAL, AND TRANSCATEGORIAL TERMS

In what follows, I shall talk pretty much interchangeably of *terms* and *concepts*, except when it is important to distinguish between constituents of language and constituents of thought. For much of the time, however, it will be more convenient to speak of *terms*, as these are obviously more immediately identifiable, being words or phrases occurring in natural or formal languages. *Sortal* terms – a locution coined by John Locke ([1690] 1975, III, iii, §15) – are nouns or noun phrases denoting putative *sorts* or *kinds* of objects (see also Strawson 1959, 168, and Wiggins 2001). They are also sometimes called *substantival general terms* (Geach 1980). Familiar examples would be the terms 'cat', 'table', and 'mountain'. They differ from *adjectival* general terms, such as 'white', 'square', and 'steep', in having not only criteria of application but also criteria of identity associated with their use. This is reflected in the fact that such terms are *count nouns*, not merely in the purely grammatical sense, but also in the more robust sense that there are principles governing their correct use in counting or enumerating objects to which they apply. A *criterion of application* tells us to which objects a general term applies and thereby fixes its extension. A *criterion of identity* tells us what conditions need to be satisfied by objects to which a general term applies if those objects are to be identical with one another. Since objects can be counted only if some principle is supplied determining whether or not certain objects to be included in the count are identical with or distinct from one another, criteria of identity are presupposed by principles for counting. For example, an instruction to count the cats living in someone's house only makes sense given a principled way to determine whether a cat encountered at one time and place in the house is, or is not, the *same* cat as a cat encountered at another time and place in the house. Clearly, different criteria of identity and principles for counting apply to objects of different sorts – for instance, to cats as opposed to mountains. Equally clearly, there are no criteria of identity or principles for counting that apply to *white* objects, say, or to *square* objects, purely insofar as those objects are white or square. After all, both cats and mountains can be white, but when we count white cats we apply different principles from those we apply when we count white mountains, and this reflects a difference between the criteria of identity associated with the sortal terms 'cat' and 'mountain'.

Sortal terms fall into *hierarchies of subsumption*. For instance, 'cat' is subsumed by 'mammal', which is in turn subsumed by 'vertebrate'. Equally, 'cat' itself subsumes, for instance, 'Siamese cat'. Every sortal term within a given hierarchy of subsumption is necessarily governed by the same criterion of identity. At the top of any given hierarchy of subsumption is a *categorial* term: in the case of the hierarchy to which 'cat' belongs, the term in question is 'animal' (or, perhaps, 'living organism'). This is the highest term in the hierarchy which shares the same criterion of identity of all the sortal terms below it in the hierarchy. Any general term that has a still more general criterion of application than this categorial term, in that it also applies to objects describable by sortal terms belonging to other hierarchies of subsumption, is a *transcategorial* term. Thus, for example, 'material object' is a transcategorial term because it applies to objects such as cats, but also to objects such as mountains, even though the sortal terms 'cat' and 'mountain' belong to different hierarchies of subsumption. Because a transcategorial term does not belong to any single hierarchy of subsumption, there can be no specific criterion of identity, or principle for counting, associated with its use. As we shall shortly see, it is important not to confuse the criterion of identity associated with a sortal term with another kind of principle governing such a term, namely, one specifying the *sortal persistence conditions* of objects to which the sortal term applies. These are the conditions that an object must continue to satisfy if the sortal term is to continue to be applicable to it, and these conditions can very often differ for different sortal terms within the same hierarchy of subsumption, or subsumed by the same categorial term. Thus, for instance, the persistence conditions of cats cannot simply be identified with those of mammals in general and, equally evidently, cats and frogs, say, have different persistence conditions, even though 'cat' and 'frog' are both subsumed by the categorial term 'animal'. It is only because all animals share the same criterion of identity that we can make sense of narratives, whether fictional or scientific, in which an animal of one sort supposedly undergoes metamorphosis into an animal of another sort while remaining one and the same individual animal. And it is because objects that are not describable by the same categorial term – such as cats and mountains – do *not* share the same criterion of identity that we cannot make sense of narratives in which, for example, an individual cat survives transmutation into a mountain. A cat could conceivably be replaced by a *cat-shaped* mountain, but the two could not conceivably be one and the same object: the cat could not continue to exist 'as' a mountain.

2   CRITERIA OF IDENTITY AND SORTAL PERSISTENCE CONDITIONS

Now something more precise needs to be said concerning criteria of identity and sortal persistence conditions. A criterion of identity is a principle expressing a non-trivial logically necessary and sufficient condition for the identity of objects of a given sort or kind, φ (Lowe 1989 and 2009, 16–28). Formally, such a principle may be stated as follows:

$$(CI\varphi) \ \forall x \forall y [(\varphi x \ \& \ \varphi y) \rightarrow (x = y \leftrightarrow R_\varphi xy)].$$

Or, in plain English: for any objects x and y, if x and y are φs, then x is identical with y if and only if x stands to y in the relation R_φ. To avoid triviality, we must insist that R_φ – which may be called the *criterial* relation for φs – is not simply the relation of identity itself. R_φ must, of course, be an equivalence relation defined on objects of the sort or kind φ – that is to say, it must be reflexive, symmetrical, and transitive and either hold or fail to hold between any pair of objects of the sort or kind φ. A paradigm example of a criterion of identity is provided by the axiom of extensionality of set theory, according to which if x and y are *sets*, then x is identical with y if and only if x and y *have the same members* – so that sameness of membership is the criterial relation for sets. But sets, of course, are *abstract* objects. It is rather harder to provide completely uncontentious examples of criteria of identity for concrete objects of any sort. This should not surprise us, since our grasp of such criteria is typically implicit rather than explicit and the criteria themselves are often open to question and revision in the light of philosophical argument. The criteria implicit in the everyday use of sortal terms are, moreover, often somewhat vague and shifting. This would be a defect in a formal language, such as the language of mathematics, but can hardly be complained about where everyday discourse is concerned.

Vagueness in our everyday criteria of identity has the consequence that some everyday questions of identity lack determinate answers, but the vast majority do not. Consider, for instance, the everyday criterion of identity for *mountains*, which is plausibly something like this:

(CI_M) For any objects x and y, if x and y are mountains, then x is identical with y if and only if x and y have the same peak.

Here we are taking mountains to be regions of terrain that are elevated above their surroundings and which possess a peak – that is, a highest point. This, no doubt, is a rather rough-and-ready definition that professional geographers might take issue with, but it will serve for purposes of illustration. Clearly, however, (CI_M) is incapable of resolving some questions of mountain identity. For instance, if we have a region of terrain that is elevated above its surroundings but in which two approximately equally high points are unsurpassed by any other, with a saddle-shaped dip between them, (CI_M) doesn't really help us to decide whether what we have here is a single mountain or two mountains separated by a shallow valley. But this doesn't mean that (CI_M) is worthless as a criterion of identity for mountains, since in the vast majority of cases it does supply a determinate answer to questions of mountain identity. The same lesson may be drawn by reflecting on the everyday criterion of identity for *animals*, which – to echo Locke ([1690] 1975, II, xxvii, §4) – is plausibly something like the following:

(CI_A) For any objects x and y, if x and y are animals, then x is identical with y if and only if x and y participate in the same life.

It may sometimes be hard to determine whether we have a case of two animals that are vitally connected to one another – as in a case of conjoined twins, or in a case of a mother and her unborn child – or just a single animal. And this is because the notion of 'sameness of life' is to a certain extent vague. But other cases are clear-cut: for instance, a rat and a flea that lives in its fur are clearly two distinct animals according to (CI_A), which is as it should be.

A further lesson that (CI_M) and (CI_A) serve to reinforce is a point mentioned earlier, namely, that objects belonging to sorts that are governed by different criteria of identity cannot intelligibly be identified with one another, with the consequence that one cannot intelligibly suppose that – to use again our earlier example – a cat could survive a process of metamorphosis which left it existing 'as' a mountain. This is because cats, being animals, have their identity determined by the relation of *sameness of* life, but mountains are simply not living things and consequently cannot be identified with anything that is essentially alive. Here it may be objected that I am just assuming without argument that, indeed, any animal is *essentially* an animal and so essentially alive. I confess that I am indeed making this assumption, though it seems an entirely reasonable one. To reject it is to suppose, in effect, that 'animal' is not, after all, a categorial term. To suppose that an animal could survive being changed into a mountain is to suppose that the sortal terms 'animal' and 'mountain' are both subsumed by some single higher-level categorial term which supplies a common criterion of identity for both animals and mountains. But what could this putative categorial term be? It could not be 'material object', since that is pretty clearly transcategorial and supplies no single criterion of identity for any of the objects to which it applies. The term applies, after all, to anything that is both an object and composed of matter. But, it seems clear, there is no single criterion of identity governing all such objects – bearing in mind here that a genuine criterion of identity must supply a *non-trivial* criterial relation for the objects that it governs.

If this claim – that 'material object' is a transcategorial term – is not immediately obvious to some philosophers, I suspect it is because they may be prone to confuse it with another term which is, pretty clearly, a categorial term, namely, 'hunk of matter'. A hunk of matter – or what Locke called a 'body', 'mass', or 'parcel' of matter – is a quantity of matter collected into a cohesive whole, which can remain intact under the impression of an external force and undergo motion as a result. Borrowing from Locke's account ([1690] 1975, II, xxvii, §3), it is not too difficult to state a plausible criterion of identity for hunks of matter, as follows:

(CI$_H$) For any objects x and y, if x and y are hunks of matter, then x is identical with y if and only if x and y are composed of the same material particles bonded together.

(CI$_H$) has the plausible implication that if some material particles are removed from a certain hunk of matter, what remains is a *different* hunk of matter. But again we should acknowledge a certain amount of harmless vagueness in (CI$_H$), arising from the fact that it is not always perfectly clear whether or not a certain 'material particle' (itself a somewhat vague term) is 'bonded' to others in a certain hunk of matter with sufficient cohesion to qualify as being a material *part* of that hunk.

This much, however, is certainly clear: that no *animal* is to be identified with any *hunk of matter*, nor is any *mountain* to be so identified, despite the fact that animals and mountains are both material objects. The truth, rather, is that, at any time at which it exists, an animal *coincides* with a certain hunk of matter, as does a mountain. Thus, if Oscar is a certain cat existing now, then Oscar now coincides with a certain hunk of matter: but Oscar is not *identical* with that hunk, because if some material particles are removed from it a different hunk of matter will then coincide with Oscar, but Oscar will remain the same cat, provided that the removal of those particles does not terminate Oscar's life. Similarly, Mount Everest presently coincides with a certain very large hunk of matter. But if that hunk of matter were to be transported intact to Australia, this would not be a way of moving Mount Everest to Australia. Rather, it would be a way of *destroying* Mount Everest, since the removal process would have left the Himalayas without that elevated region of terrain that is Mount Everest.

So far in this section, I have said a good deal about criteria of identity, but nothing yet about sortal persistence conditions. Sortal persistence conditions are the conditions that an object of a given sort must comply with in order to continue to exist *as an object of that sort*. Clearly, the criterion of identity governing a given sort of objects must be complied with by any such object if it is to continue to exist *at all*, but this is not enough for its persistence as an object *of that sort*, since its compliance with the criterion of identity is compatible with its 'metamorphosis' into an object of another sort governed by the same criterion. A simple example to illustrate this point is the following. Mount Everest is, obviously, currently a mountain: but if sea levels were to rise dramatically, it could be transformed into an *island*. Mountains and islands are different sorts of 'geographical prominence' – to coin a word for the categorial term applicable to them both – since an island is necessarily surrounded by water, whereas a mountain (a terrestrial mountain, at any rate, as opposed to an undersea one) is necessarily *not*. A slightly more controversial example involving animals is provided by the case of a caterpillar and the butterfly into which it is eventually transformed. It is apparently the case that, in the chrysalis stage, the internal structure of the caterpillar is completely destroyed and reorganized, making this unlike the simple development of an amphibian, for example, from its larval to its adult phase. Nonetheless, throughout the metamorphosis of the caterpillar into the butterfly, the same *life* continues, making the butterfly the same animal as the caterpillar, even if it is not, in the relevant sense, the same *sort* of animal. This example is, of course, a scientific one which has a basis in empirical fact. But fictional examples of a similar kind are familiar from folklore, as in the story of the frog prince, in which a human being is transformed into an amphibian. We can make sense of this story, even while dismissing it as pure fiction, because we can intelligibly suppose the prince and the frog to participate in the same uninterrupted life.

Sortal persistence conditions, to the extent that they go beyond the demands of criteria of identity, are only discoverable by empirical means. But in order to discover those conditions empirically, we must first be able to grasp the relevant criteria of identity. For example, it is only because we already know or assume that sameness of life is the criterial relation for animal identity, that we can then go on to determine whether, when a caterpillar is transformed into a butterfly, this is to be classified as an individual's surviving a change of *sort* or merely as its surviving a change of *phase* within the same sort. Consequently, criteria of identity, although they obviously do not lack empirical content, have the status of 'framework principles' rather than mere empirical discoveries. Revising or amending a criterion of identity is, thus, more in the nature of a methodological or indeed a philosophical exercise than is revising or amending an account of the sortal persistence conditions governing a sort or kind, at least where natural kinds are concerned. A great deal more can and should be said about such matters, but enough has now been said for the purposes of the present chapter.

3   TWO NOTIONS OF INDIVIDUATION

At the beginning of this chapter, I said that I would be defending a certain thesis regarding the individuation of objects that I call *categorialism*. But to make this thesis clear, it is first necessary to distinguish between two different, albeit related, notions of *individuation*. The notion of individuation with which I am chiefly concerned at present is a purely *cognitive* notion. Individuation in this sense is a kind of cognitive *achievement*, namely, *the successful singling out of an object in thought*. Categorialism is then the doctrine that a thinker can single out an object in this way only if he or she grasps, at least implicitly, some categorial concept under which he or she conceives the object in question to fall. When a thinker thus singles out an object in thought and has a specific thought about *that very object*, we have a case of singular, or *de re*, thought about that object. And it seems clear that at least sometimes thinkers do have such singular thoughts. I shall have much more to say about such thoughts shortly. Before that, however, I need to say something about the other notion of individuation.

In addition to the notion of individuation in the cognitive sense, we have the notion of individuation in the *metaphysical* sense. Individuation in this sense has nothing to do with thinkers or thoughts – except, of course, when we are talking about the metaphysical individuation of thinkers and thoughts themselves, for these, like objects of any other kind, are subject to principles of individuation. A principle of individuation, in the metaphysical sense, is a principle telling us how objects of a certain sort are singled out *in reality*, as opposed to how they are singled out *in thought*. Such a principle tells us what makes an object of a certain sort the very individual that it is, as opposed to any other individual of the same or indeed of any other sort. Unsurprisingly, then, a principle of individuation is closely related to a criterion of identity, although the two notions do not perfectly coincide. A criterion of identity for objects of a sort φ tells us what it takes for such objects to be the same φ or different φs. But it doesn't necessarily tell us what makes such an object the particular φ that it is. For example, our criterion of identity for mountains, (CI_M), tells us that mountain A is the same mountain as mountain B if and only if A and B have the same peak. But it doesn't appear that a particular mountain, such as Mount Everest, is *individuated* by its peak – not, at least, by its peak alone. To know *which mountain* Mount Everest is, we need to know not only which peak is its peak, but also, and more importantly, its geographical extent. Only when we know that do we know where Mount Everest 'begins' and where other mountains in the region 'leave off'. That is to say, only then do we know how Mount Everest is singled out in reality from other surrounding mountains, and indeed from the air moving about it and the animals living on its slopes.

I have just spoken of knowing *which* mountain Mount Everest is. Someone who does know this can certainly individuate Mount Everest in the cognitive sense – can single that mountain out in thought and have singular thoughts about that very mountain. So, a thinker's knowledge of the principle of individuation governing an object, which is intimately bound up with its criterion of identity, is clearly at least *part* of a sufficient condition for that thinker's being able to single out that object in thought. The example of sets may provide a useful illustration once more. A set, it seems clear, is individuated – in the metaphysical sense – by its members, so that in this case, at least, there is a coincidence between the criterion of identity for sets and their principle of individuation. However, in order to know *which* set a given set is, it does not suffice merely to know that this set, like any other set, is individuated by its members. What *does* suffice is to know not only this but also *which objects* are in fact the members of the set in question. We may call a set's members its *individuators*. And the point of the example is to show that a *sufficient* condition for a thinker's being able to single out an object in thought is his or her knowing, not only the object's principle of individuation, but also its individuators.

Mere knowledge of the object's individuators does not suffice without knowledge that *they are* its individuators, which requires knowledge of the object's principle of individuation. A child might know which objects are the members of a certain set – for instance, that two, three, five, and seven are the members of the set of prime numbers smaller than ten – but without its knowing that these numbers are what individuate that set, it is doubtful whether the child could really be said to know *which* set that set is, since it might be under the mistaken impression that another set could have these same members.

However, I want to make it quite clear that I am not presuming at this point that a thinker can single out an object in thought *only if* he or she knows *which* object it is and knows this by knowing not only its principle of individuation but also what its individuators are. I am merely saying that knowledge of the latter kind *suffices* for a thinker's ability to single out an object in thought. I also believe that thinkers certainly do sometimes have knowledge of this kind and are, in virtue of it, able to single out certain objects in thought. The question, however, is whether such knowledge is *necessary* for that achievement and, if not, what condition *is* necessary. Categorialism maintains that *one* necessary condition is the thinker's grasp – which may only be implicit – of a categorial concept under which he or she conceives the object in question to fall.

4   OBJECT PERCEPTION AND SINGULAR THOUGHT

The case discussed in the previous section, of a thinker's singling out in thought a particular set – the set of prime numbers smaller than ten – is special in that the object in question is an *abstract* object, and hence not one that can be perceived by the senses. It may well be supposed that singling out in thought an abstract object like this is particularly demanding intellectually, for precisely this reason. By the same token, it may be supposed that if an object can be *perceived* by a thinker, the thinker's perception of the object provides him or her with a way of singling it out in thought which is intellectually much less demanding – so much so that in cases like this categorialism is a most implausible doctrine. Of course, this presumes that categorialism does not extend beyond thought to perception itself: that is, it presumes that it is false to maintain that a subject cannot even *perceive* an object without grasping some categorial concept under which he or she conceives the object in question to fall. But *perceptual* categorialism, as we may call this more extreme doctrine, is a most implausible thesis. It would imply, for instance, that animals lacking categorial concepts – as, quite plausibly, many higher mammals do – cannot perceive objects in their environment.

For my own part, I am content to deny perceptual categorialism. I favour a *causal* theory of object perception according to which – to a first approximation – a subject S perceives an object O if and only if certain features of S's perceptual experience, discriminable by S, are differentially causally dependent on certain properties of O (Lowe 1996, 102–117). A feature F of S's experience is *differentially* causally dependent on a property P of O just in case variations in P would cause corresponding variations in F. So, for example, by this account I currently *see* a brown wooden table standing in front of me if, say, variations in its shape, orientation, or hue would give rise to corresponding variations in certain features of my current visual experience, discriminable by me. If variations in properties of the table would give rise to *no* variations in my current visual experience, or only to variations indiscriminable by me, then, I think, I do *not* see the table. This account of object perception imposes no direct epistemic requirement on such perception: a subject is not required, by this account, to know *what* object he or she perceives, nor even *that* he or she is perceiving an object. At most the account defines object perception in such a way that a subject's perception of an object makes it *possible* for the subject to know what he or she is perceiving, and that he or she is perceiving it, provided that he or she meets certain other requirements for such knowledge, yet to be specified. For instance, it might be insisted, as a minimum requirement, that a subject cannot know what he or she is perceiving, or that he or she is perceiving it, unless he or she is *attending* to the object in question. I should also make it clear that, in saying that the relevant variations in the subject's experience should be *discriminable* by the subject, I am not smuggling in an epistemic requirement on object perception. For I do not require the subject to be able to form *discriminatory judgements* regarding features of his or her perceptual experience – only to be *sensitive* to relevant variations in his or her perceptual experience. A subject could be tested for such sensitivity without being asked whether or not he or she could notice any experiential difference of a relevant kind. This is still only a rather sketchy account of object perception, but it will suffice for the discussion that is to follow.

Categorialism is likely to be challenged on the basis of examples such as the following. A subject, S – perhaps an infant or a young child – is confronted with an object, O, which S sees. Perhaps, furthermore, S directs his or her *attention* towards O and, as O moves, S *tracks* O's movement with his or her eyes, continuing to direct his or her attention towards O.

Now let the question be asked: can S, in virtue of this sort of perceptual connection with O, have singular thoughts specifically *about* O – thoughts about O such as that *it is white* or that *it is furry*? Of course, we must presume that S can indeed *think* and, moreover, can have thoughts of the *form* 'X is white' and 'X is furry.' Our question is only whether, in order to have such singular thoughts about O in particular, all that is necessary in addition to these general cognitive capacities is that S should have the kind of perceptual connection with O that has just been described. A good many theorists would, I am sure, answer this question positively. But a *categorialist* would not. A categorialist will insist that, unless S conceives of O in a certain way, even if only implicitly, and is broadly *correct* in so conceiving of O, S's perceptual connection with O does not enable S to have singular thoughts about O. The conceptual requirement in question is this: S must, at least implicitly, conceive of O as falling under a certain categorial concept. This requirement will be met if, say, S conceives of O as being an *animal.* Thus, if S is perceptually connected with O in the described fashion, S conceives of O as being an animal, and O *is* an animal, then S is thereby put in a position to have singular thoughts about O. But it is not necessary that S should *explicitly* conceive of O as being an animal. It suffices that S should conceive of O as being some specific *sort* of animal, such as a *cat.* And in that case it is not necessary that O should actually *be* a cat, only that it be *some* kind of animal. Thus, S can be mistaken about what sort of animal O is and still be put in a position to have singular thoughts about O. But, according to categorialism, S cannot be put in a position to have singular thoughts about O, even if S is perceptually connected with O in the described fashion, if S either fails, even implicitly, to conceive of O as falling under any categorial concept at all or else misconceives the putative object of his or her perception as being something falling under a categorial concept which does not apply to O.

It seems to me that this contention is susceptible of *proof,* or at least to the nearest thing to proof that is normally available in questions of philosophy. Let us suppose that O is, in fact, a *cat.* That is to say, let us suppose that S is perceptually connected to a certain cat in the described fashion: S *sees* the cat, S directs his or her *attention* towards the cat, and S *tracks* the cat's movements with his or her eyes, continuing to direct his or her attention towards the cat. Call the cat in question 'Oscar'. But let us also suppose that S does not conceive of Oscar as *being a cat,* nor indeed as falling under *any* sortal or categorial concept. Then the anti-categorialist is presented with the following difficulty. As we have already observed, whenever a certain animal exists in a certain place at a certain time, it coincides with a certain *hunk of matter*, which is numerically distinct from the animal in question. Now, the perceptual connection that S has with respect to Oscar, the cat, S will very likely *also* have with respect to a certain hunk of matter – namely, the hunk of matter composing Oscar during the perceptual episode in question. I say 'very likely' because, of course, it is at least possible that *different* hunks of matter should compose Oscar at the beginning and end of the episode – if, for example, Oscar should lose some hair or ingest some food during the course of that episode. However, let us set aside this possibility for the time being and return to it later, since it will turn out to be an important one.

The problem for the anti-categorialist is now to explain why S's perceptual episode should put S in a position to have singular thoughts specifically about *Oscar*, as opposed to the hunk of matter – call it 'Hunk' – that, as we are now supposing, composes Oscar throughout the episode. If the anti-categorialist maintains, in the light of this challenge, that S is put in a position to have singular thoughts about *both* Oscar *and* Hunk, then another problem arises: what makes one of those thoughts specifically a thought about *Oscar* and another specifically a thought about *Hunk*? A genuinely singular, or *de re*, thought must determinately be a thought about a particular, uniquely identifiable object. I am not insisting that the *thinker* of the thought must be able to *identify* that object – to do so would clearly be question-begging in the present context – only that there must be *some* determining factor which makes the thought in question a thought about the particular object in question. But, as far as I can see, the anti-categorialist has no resources with which to say what this determining factor could be, since in a case such as that just sketched two *different* objects are symmetrically related to the thinker in the only way that the anti-categorialist requires for singular thought about them. By contrast, the categorialist has a tiebreaker for this kind of situation. According to categorialism, what makes *Oscar* the object of one of S's thoughts is the fact that S conceives of one of the objects being perceived as being a *cat* – or at least as being an *animal* or some *sort* of animal – and is correct in doing so.

A discussion is now needed of the more complicated case in which, say, Oscar loses some hair or ingests some food as it moves through S's field of view. If S continues to track *Oscar*, can't the anti-categorialist maintain that *this* is why S is put in a position to have singular thoughts specifically about Oscar? For, of course, in this case there is no *single hunk of matter* that S is tracking. Moreover, the anti-categorialist may now go on to claim that the original example, in which Oscar does not change in material composition, only presents a special case of this more complicated one.

The anti-categorialist may urge that what puts S in a position to have singular thoughts specifically about Oscar in *all* such cases, complicated or simple, is a fact expressible by the *subjunctive conditional* that S *would* continue to track Oscar *if* Oscar were noticeably to lose or gain some appropriate component matter, such as some hair or some food. However, far from this suggestion serving to aid the anti-categorialist cause, I think that it effectively concedes victory to categorialism. For recall that I insisted only that S should at least *implicitly* conceive of the object of perception – in this case, Oscar – as falling under some categorial concept which does in fact apply to that object, namely, *animal*. I don't require S to be able to *articulate* this concept or its associated criterion of identity explicitly. But if S is demonstrably capable of perceptually tracking a particular animal through various changes to its material composition which are specifically permitted by the criterion of identity for *animals*, but *not* by the criteria of identity governing other categorial concepts, including that of *hunk of matter*, then this provides strong evidence that S *does* in fact possess an implicit grasp of the categorial concept of an animal. And bear in mind here that not just *any* change of material composition is compatible with the continuing existence of an animal.

Indeed, we can now see more clearly why it is *categorial* concepts that are crucial to the capacity for singular thought about objects encountered in perception, for it is precisely these concepts that are most intimately tied to *criteria of identity*. In cases in which two initially coinciding objects, such as a cat and a hunk of matter, go their separate ways in the course of some extended perceptual episode, the *correct* way to track those different objects will be determined precisely by their respective criteria of identity – whence it is reasonable to conclude that subjects who *succeed* in correctly tracking those objects exhibit at least an implicit grasp of the criteria of identity governing those objects, and thereby an implicit grasp of the categorial concepts under which they fall and which determine those criteria of identity. In short, the very perceptual-tracking considerations that the anti-categorialist is prone to regard as *defeating* the claims of categorialism in fact serve to *support* those claims, when these considerations are given their proper scope.

What if it should turn out that S – who, remember, we have supposed to be an infant or young child – doesn't consistently and comprehensively track Oscar in *all* kinds of circumstances, in accordance with the criterion of identity for animals, but still clearly does so in certain limited kinds of circumstances? Suppose, for instance, that S tracks Oscar successfully through the loss of some hair and the ingestion or excretion of food matter, or through changes in the dispositions of Oscar's limbs and tail, but that S fails to do so through periods in which Oscar is curled up and asleep. Then I think we should say that S hasn't *fully* grasped, even only implicitly, the *adult* concept of an animal and its associated criterion of identity, perhaps because S doesn't yet understand that an animal can still be alive when it is asleep. But I still think that S is evincing the implicit grasp of *some* categorial concept with an associated criterion of identity – a concept which we might describe as being an infantile precursor to the fully fledged adult concept of an animal (a precursor concept which might well be *innate* in humans). And we may deem that this grasp is sufficient for S to have singular thoughts about Oscar, at least with respect to the limited kinds of circumstances in which the infantile criterion does not significantly come apart from the adult one in terms of its implications for the identity of the tracked object. A suitably nuanced version of categorialism should certainly be prepared to admit such qualifications as these to its basic doctrine.

5 CATEGORIALISM AND LINGUISTIC REFERENCE

From this point forward, I am going to assume that categorialism is correct. But categorialism is a thesis about *singular thought*, not about *linguistic reference*. Even so, it may – indeed, I think it *does* – have important implications for linguistic reference. According currently dominant 'causal', or 'direct', or 'anti-descriptivist' theories of reference, all that is required for a speaker's use of a proper name or natural-kind term, N, to be referentially successful – that is, for the speaker to use N to refer successfully to N's referent – is that the speaker's use of N should stand in the right kind of causal-historical relation to a 'reference-fixing' event in which N was first introduced into the speaker's speech community (Kripke 1980). The 'right kind' of causal-historical relation consists, supposedly, in a chain of usage acquisition, where each speaker S in the chain – apart, of course, from the first – uses N with an intention to refer to *the same thing* as did the speaker S' from whom S acquired the use of N. The first speaker in the chain is the person who first introduced N into the speech community. According to this sort of account – which, to be fair, has only been sketched in a very bare-bones way here – successive speakers in the chain of acquisition do not need to associate with N any *descriptive information* which applies to N's referent in order to be able to use N successfully to refer to that referent. Hence, in particular, it is not the case, on this view, that the referent of N, as used by a given speaker, is determined by a certain *definite description* which the speaker associates with N.

Now, I am happy to concede that there may be a relatively attenuated and anodyne notion of 'reference' for which this sort of account is broadly acceptable. But if categorialism is a correct doctrine concerning singular thought, such an account of linguistic reference cannot be the whole story. This is because we must acknowledge that, at least sometimes, speakers use sentences containing proper names and natural-kind terms to express *singular thoughts*. Suppose that some friends of my neighbours have acquired a new pet cat, which they decide to call 'Oscar'. And suppose that I overhear a conversation between my neighbours and their friends during the course of which they say, among other things, that Oscar is white and beautiful, but never say anything that reveals, either explicitly or implicitly, that Oscar is *a cat*. According to the causal-historical theory of linguistic reference, I am now in a position to refer successfully to *Oscar*, provided I use the name 'Oscar' with the intention to refer to whatever it was that my neighbours and their friends were referring to when they used the name in the conversation that I overheard. The *only* constraint on my referential success with the name 'Oscar' is supposed to be that I should intend to use it to refer to *the same thing* that my neighbours and their friends were using 'Oscar' to refer to. But 'thing' is not a categorial term: it is a *transcategorial* term, like 'object'. It carries with it no specific *criterion of identity*. Consequently, this supposed 'constraint' on my successful use of the name 'Oscar' is very slim indeed. As far as I am concerned, Oscar might be almost *any kind of thing whatever*. Indeed, if I were honest I would have to confess that I have *no idea* what kind of thing Oscar is. Now, as I say, I am relatively content to acknowledge that there is *some* sense in which I am referring to Oscar, even when I say that I have no idea what kind of thing Oscar is. But the more important question, I think, is whether I can intelligibly be said to be able to use the name 'Oscar' in sentences to express *singular thoughts* about Oscar. I don't see how I can, given that categorialism is a correct doctrine concerning singular thought. For categorialism requires that, in order for me to have a singular thought about Oscar, I must at least conceive of Oscar, correctly, as falling under some categorial concept – so that, since Oscar is in fact a cat, I must at least conceive of Oscar as being an *animal* of some sort. But in the envisaged scenario I do not conceive of Oscar in this way, nor do I conceive of Oscar as falling under any other categorial concept.

Here it may be urged that this just shows that categorialism must be *wrong*, on the grounds that I can at least use the sentence 'I have no idea what kind of thing Oscar is' to express a singular thought of mine about Oscar. But that response is question-begging in the present context, given that I have just spent a good deal of time *defending* categorialism. Merely to *assert* that such a sentence may be used to express a singular thought about Oscar is no rebuttal of my previous arguments. At most it can be demanded of me that I give a plausible account of what sort of thought such a sentence *could* be supposed to convey, assuming the truth of categorialism. But this is surely not difficult. The obvious answer, which I think is a plausible one, is that 'I have no idea what kind of thing Oscar is' is implicitly *metalinguistic*. The thought that I am conveying by it is the thought that *I have no idea what kind of thing the referent of the name 'Oscar' is*. Here it might be objected that this only postpones the problem, because I have just replaced the name 'Oscar' by the singular noun phrase 'the referent of the name "Oscar"'. But, of course, even by the lights of the causal-historical theory of reference, this definite description should not be assimilated to a proper name. Rather, we can explain its proper use in terms of something like Russell's theory of descriptions, that is, in *quantificational* terms. Indeed, another way of expressing the thought conveyed by my use of the sentence 'I have no idea what kind of thing Oscar is' would be something like this: *the name 'Oscar' refers to something, but I've no idea what kind of thing it is*. And, of course, the 'it' in 'I've no idea what kind of thing it is' is not playing a *referential* role, but rather the role of a *variable of quantification*, bound by the preceding quantifier 'something'.

So, my contention is that, when a speaker uses a sentence containing a proper name or natural-kind term to express a singular thought about the referent of that name or natural-kind term, he or she must at least possess implicit knowledge of what *categorial* concept the referent falls under. This knowledge will obviously be *descriptive* knowledge, of a certain kind, about the referent. But in saying this I am not simply returning to the 'descriptive theory of reference', which causal-historical accounts of reference sought to overturn. For I am not contending that the descriptive knowledge of the referent that the speaker must possess in order to use the proper name or natural-kind term in sentences expressing singular thoughts about the referent must be descriptive knowledge which determines precisely *which* thing is that referent. Nor, on the other hand, am I arguing against the latter view. Rather, at this point I would rather keep an open mind about the matter.

Another important application of categorialism to the theory of linguistic reference concerns the notion of a 'reference-fixing event', of the kind involved, according to the causal-historical account of reference, when a proper name or natural-kind term is first introduced into a speech community. Let us return to the story of my neighbours' friends' pet cat.

It was said that the friends decided to call their new pet cat 'Oscar'. But to do that, they needed to have *singular thoughts about Oscar*. It may be that they first *saw* Oscar in a pet shop and, looking at him, decided there and then to call him 'Oscar'. Perhaps one of the friends said to the other, 'Let's buy him and call him "Oscar".' But this required the friend to use the pronoun 'him' in a sentence expressing a singular thought about *the cat in front of them*. Consequently, according to categorialism, this friend had to conceive of that cat at least as being an *animal* of some sort.

What, though, if these people were seriously mistaken about the nature of the shop they had entered. Suppose they thought it was a china shop and that what they were looking at was a *china ornament* of a cat (perhaps Oscar was fast asleep at the time). Wouldn't they still have succeeded in having singular thoughts about *that cat* and have succeeded in naming him 'Oscar'? It might appear that common sense is on the side of those who answer this question positively, but we need to think the matter through more carefully. What entitles us to assume that what the friends would have succeeded in deciding to call 'Oscar' was *the cat* that they were looking at, given that they didn't conceive anything that they were looking at to *be* a cat? It won't do to answer here that the thing that they *were* looking at was in fact a cat, because this presumes that there was just *one* thing that they were looking at. But, as was remarked in an earlier section, whenever an *animal* of any sort is located at a certain place at a certain time, it coincides with a certain *hunk of matter*, which is distinct from the animal in question. So there were – at least – *two* things that the friends were looking at, a *cat* and a *hunk of matter* coinciding with that cat. Why, then, should we suppose that, despite their mistake, they succeeded in calling *the cat* 'Oscar' rather than *the hunk of matter*? To this it may be replied that they certainly didn't intend to be naming a *hunk of matter*. So the suggestion is that, by default, what they succeeded in naming was *the cat*. But that suggestion is question-begging in the present context, for it assumes that they did indeed succeed in naming *something* – and yet whether they did so is precisely the question now at issue. In any case, how can something be named *by default* in this supposed fashion? How can an intention *not* to name one thing somehow bring it about that *another* thing is named?

The suggestion that, in the scenario just described, my neighbours' friends succeeded in naming *a cat* 'Oscar', despite being under the impression that they were naming a *china ornament*, also faces the following difficulty. Suppose that the friends never discovered their mistake, because they were suddenly called away and never managed to return to the shop.

And suppose that, later, reminiscing with my neighbours about the incident and expressing their regret at not having been able to make their purchase, they passed the name 'Oscar' on to my neighbours. And suppose that my neighbours also believed that 'Oscar' was the name of a certain china ornament that their friends had intended to purchase. According to preceding arguments of this section, my neighbours are not in a position to express singular thoughts about *the cat* that their friends actually saw, when using sentences containing the name 'Oscar'. And that seems right. If they say such things as 'Oscar would probably have been broken by now, because our friends are so clumsy with ornaments', it surely makes no sense to suppose that they are expressing such thoughts about a certain *cat*. But if that is the right thing to say about my neighbours' use of the name 'Oscar' – that it simply *fails to refer* – then it is surely also the right thing to say about their friends' use of it, even though it was they who introduced the name in the first place. Certainly, this is the verdict of categorialism and I consider it to be a virtue, not a failing, of the doctrine that this is so.

Finally, I should note that although I have been concentrating on the reference of *proper names* in this section, I intend all of its lessons to apply also to *natural-kind terms*, although the adjustments to the account that would be needed to extend it to these would take up more space than the present chapter allows.

REFERENCES

Geach, P. T. (1980) *Reference and Generality*, 3rd edn. Ithaca, NY: Cornell University Press.
Kripke, S. A. (1980) *Naming and Necessity*. Oxford: Blackwell.
Locke, J. ([1690] 1975) *An Essay Concerning Human Understanding*, edited by P. H. Nidditch. Oxford: Clarendon Press.
Lowe, E. J. (1989) What is a criterion of identity? *Philosophical Quarterly* 39: 1–21.
    (1996) *Subjects of Experience*. Cambridge University Press.
    (2003) Individuation. In M. J. Loux and D. W. Zimmerman (eds.) *The Oxford Handbook of Metaphysics*. Oxford University Press, pp. 75–95.
    (2007) Sortals and the individuation of objects. *Mind & Language* 22: 514–533.
    (2009) *More Kinds of Being: A Further Study of Individuation, Identity, and the Logic of Sortal Terms*. Malden, MA, and Oxford: Wiley-Blackwell.
Strawson, P. F. (1959) *Individuals: An Essay in Descriptive Metaphysics*. London: Methuen.
Wiggins, D. (2001) *Sameness and Substance Renewed*. Cambridge University Press.

CHAPTER 7

Action, perception, and reference

Peter Machamer and Lisa Osbeck

How do perception and reference connect to each other and to realism? Obviously, these cannot be taken as problems solely in the philosophy of language – as a question of how words and their meanings relate to the world. Yet since Frege until recently, reference has pretty much been taken as a problem in philosophy of language. In perception, most simply, there are no words,[1] even if linguistic competence eventually affects what we pick out in the perceptual array. Nonetheless, there are problems about explaining the relations between perception and the world, and relations of perception to concepts and words. The problem of reference is to understand how percepts, concepts, and words are tied to the world, and how relations to the world form part of their meaning.

Always a complicated matter, the problem of reference has become even more complex because of contemporary trends in cognitive science. Not too long ago, most cognitive scientists thought that passive individualism, internal computation, and logic would provide the model to solve all problems, including the problem of reference. Logical relations were interpreted over sets of objects and their properties, which were the instantiations of predicates. Extensionality (ranging over objects and properties) was the mode of interpreting logical formulae. The focus of interpretation was the range of objects that instantiated the abstract logical-nominative or adjectival-properties description.

Of late, however, in many contexts, the logical-relations framework, and by extension the focus on objects and their importance, are increasingly called into question, if not rejected out of hand. Ideas once considered unconventional or even bizarre by the cognitive-science orthodoxy

---

[1] For example, Stephen Palmer's definition of 'visual perception' is fairly typical: 'visual perception will be defined as the process of acquiring knowledge about environmental objects and events by extracting information from the light they emit or reflect' (Palmer 1999, 5). There are many interesting and important questions about perception and language, but accounts of perception per se do not require mentioning language.

now belong to the commonplace. A wave of empirical support, persuasive arguments, and new methods converge (e.g. Brooks 1991; Clark 2003, 2008; Hutchins 1995a, 1995b; Lakoff and Johnson 1999; Suchman 2007; Tomasello 1999) so that it now appears stuffy, if not peculiar, to deny the situated and embodied nature of cognition and language or the active nature of perception and learning. The following summary of this trend appears, in all places, in *Artificial Intelligence*:

> For over fifty years in philosophy, and for perhaps fifteen in Artificial Intelligence and related disciplines, there has been a re-thinking of the nature of cognition. Instead of emphasizing formal operations on abstract symbols, this new approach focuses attention on the fact that most real-world thinking occurs in very particular (and often very complex) environments, is employed for very practical ends, and exploits the possibility of interaction with and manipulation of external props. It thereby foregrounds the fact that cognition is a highly *embodied* or *situated activity* – emphasis intentionally on all three – and suggests that thinking beings ought therefore be considered first and foremost as acting beings. (Anderson 2003, 91)

However, the extent to which these theoretical changes, and in particular the emphasis on acting beings, have been codified into a coherent view centering on the active natures of individual, environment, and group is problematic. Situated and embodied cognition (and related perspectives, such as distributed cognition and extended mind) assume a fundamental role for activity, that is, for the importance of organisms (including humans) acting in and on the world and the world's acting on organisms. Yet much remains fuzzy as to the specific means by which human activities impact on what is cognized. Cognitive development is one area that remains especially problematic. One aim of this chapter is to examine some of the ways activity, action, embodiment, and situation have impacted accounts of conceptual development, perception, and linguistics and to point out a few places where there is room for greater impact. As a preview, we claim that some prominent accounts of the acquisition of concepts remain rooted in a logical-relations framework, virtually untouched by the 'new' cognitive psychology, neuroscience, or linguistics. Along the way we compare the treatment of activity within contemporary theories of conceptual development with the ways in which activity is theorized in classical pragmatist and ecological frameworks. One fundamental problem that these frameworks and their contemporary offshoots (e.g. Brown et al. 1989) attempt to avoid is the problem of reference, by prioritizing activity. The problem of reference is often taken to be about the nature of representations; yet both pragmatist and ecological approaches sometimes eschew talk of representation altogether (e.g. Costall 1995; Lave 1988). Approaches for which activity features prominently uphold a strong distinction between representation *of* objects encountered and participation *with* objects (including other persons) through interactions or transactions of various sorts. Activity in essence is a kind of functional relation with objects, always aimed at an object in accordance with a need (Leont'ev [1959] 1981). Thus the world is always implicated in activity, and activity always implicated in cognition: 'Learning and acting are interestingly indistinct, learning being a continuous, life-long process resulting from acting in situations' (Brown et al. 1989, 33).

But can the problem of reference genuinely be 'bypassed' with a turn to activity? The language from activity theorists is always about integration and interdependency, and the lack of meaningful distinction between concepts and activity. But surely the very dichotomy traditionally upheld between representation and activity theory might be dissolved by recognizing that representing is a form of human activity fundamental to adaptive functioning. Moreover, all representation is not alike: the representation central to the 'spectator view' of knowledge that was so problematic to Dewey differs fundamentally from some contemporary theories of representation. For example, the 'new representationalism' of Prinz (1992, 2005), Decety (2001), and others argues for a 'common coding' at the neural level for perception of movement and its execution, thus eliminating the need for further processing or transformation between the perceptual and activity levels. The idea is that perception and action are interdependent and *coactivated*, which constitutes a departure from the older view that perception *causes* action. That this 'double-layer model – supporting both representation and non-representational activation' is compatible with contemporary views of cognition as situated is emphasized in a recent paper by Chandrasekharan and Osbeck (2010). As they claim, the 'ideomotor' principle implicated in the common-coding view is anticipated in important respects by William James (1890), who noted the 'awakening' of movement in the body through the representation of another's movement (526). More recently, Brass and Heyes (2005) review the extensive literature on imitation and suggest that imitation of movement might well be achieved by the automatic activation of existing motor representations through the observation of movement – suggesting that watching a performance is more effective than listening to instructions in learning new skills. Much support comes from neuroimaging studies. For example, Gallese et al. (2002) found that identical cortical areas are activated when observers watch others engage in various goal-directed activities. The effect appears to be strengthened with expertise. Expert dancers show strong activation in parietal, premotor, and posterior STS (superior temporal sulcus) regions when watching a performance of their preferred dance form; non-dancers do not show these effects. There is also evidence that imagining activity results in activation of relevant brain areas and in some cases, explicit behavior that replicates the imagined movement, such as sharpshooting (Barsalou 1999). These effects have been extended to language processing, in studies demonstrating motor activation in participants imagining or processing words referencing movement (Barsalou 1999; Holt and Beilock 2006; Wilson and Gibbs 2007).

These and similar effects are explained in terms of the activation of mirror neurons (Fadiga et al. 2000). Mirror neurons are attuned to specific intentional motor actions performed with regard to a limited range of objects (e.g. picking up bananas, but not a glass of water), but are reasonably indifferent as regards what individual is performing the act (whether self or any others). As the discoverers put it:

In the early 1990s our research group at the University of Parma in Italy … found that answer somewhat accidentally in a surprising class of neurons in the monkey brain that fire when an individual performs simple goal-directed motor actions, such as grasping a piece of fruit. The surprising part was that these same neurons also fire when the individual sees someone else perform the same act. Because this newly discovered subset of cells seemed to directly reflect acts performed by another in the observer's brain, we named them mirror neurons. (Rizzolatti et al. 2006, 56)

The adaptive value of common coding and mirror neurons is emphasized recently, as are also organisms as activity systems. Chandrasekharan and Stewart (2007) suggest on the basis of modeling work that common coding arises both via evolutionary and developmental (within-lifetime) mechanisms as organisms shift from non-representational functioning to modes of functioning that enable the use of representations. The wider point here is that theories that prioritize human activity in development and elsewhere need not shun 'representation' out of hand but rather identify notions of representation compatible with the emphasis on adaptation to the contexts of learning. In turn, however, theories of cognitive development should consider how activities in and of the world play important roles in the human activities of representing or referring. As we will discuss below, the causal connections established by organisms acting in the world allow for a way of thinking about reference, with or without representations.

CHILD DEVELOPMENT

We begin to address this question with brief review of a few prominent contemporary accounts of conceptual development, and how activity is incorporated therein. Theories of conceptual development generally start with a description of the capacities and propensities of prelinguistic infants. Jean Piaget ([1937] 1954) is the major progenitor, but most everyone now objects to Piaget's time frames for, and stages of, conceptual development. Piaget had sensorimotor schemata as his first stage, but many recent responses emphasize that infants have conceptual representations that go well beyond perception. Nonetheless, it is still worth noting that Piaget's schemata were action schemata and, for him, action schemata provided the original basis for all the other structures that would come on-line as the child grew and experienced more. This was a sharp departure from the traditional empiricist line that only the perceptual properties of objects are available to the child, and that everything else has to be constructed from such a limited perceptual base. The action emphasis has been pushed further by many more recent developmental theorists. Let us look at two examples.

In 2004 Jean Matter Mandler published *The Foundations of Mind* (Mandler 2004), where she argued that even very young children (under twelve months) have a conceptual system, which she contrasted with Piaget's claim that such children only had sensorimotor schemata. Mandler is adamant that perceptions are not concepts; one of her aims is to draw a harder line around percepts and concepts than has been acknowledged in development, specifically around the accessibility of knowledge. She defines 'concept' as '*declarative* knowledge about object kinds and events that is potentially accessible to conscious thought' (2004, 4; emphasis added). The very idea of prelinguistic declarative knowledge is a curiosity, at least. Mandler's claim is that infant concepts consist in image schemas – spatial representations – and that these image schemas function 'to structure and give meaning to the images that we are aware of when we think' (14). They are not images, but rather schematic spatial representations. Her image schemas are like those 'idealized cognitive models' found in the work of George Lakoff and Mark Johnson (1980; Lakoff 1990), whom she credits liberally. Examples include SOURCE–PATH–GOAL, CONTAINMENT, SUPPORT, CAUSED MOTION, SELF-MOTION, UP–DOWN, and others. As such they lack the symbolic and syntactic structure by which concepts are understood by other theorists such as Carey (below). They form the foundational meanings of the conceptual system and 'allow language to be learned' (Mandler 2004, 119). Moreover, they enable the infant to engage in forms of 'theorizing' about objects, albeit primitive theorizing, even 'reflective abstraction', 'inductive inference' (115), 'interpretation' (13). Piaget reserved this ability, or at least this terminology, for the systematic mind capable of concrete operations (approximately age seven). But Mandler's view is that through their image schemas, even young infants can generalize from experience and engage in rudimentary forms of prediction about objects and analogical thinking that paves the way for linguistic competence. Mandler thus regards infants as 'much more thoughtful than they have typically been given credit for' (2004, 11), and assumes continuity between infant and adult cognition rather than a set of stages. Image schemas 'may serve the same function as words in infants' thought, but they are not static units in the way that words are' but are rather 'dynamic representations' in addition to being 'productive' (79) and 'combinatory' (91). Much of the evidence comes from studies interpreting spontaneous forms of play with objects as demonstrating problem-solving strategies and imagination, as when a ten-month-old accustomed to playing 'peekaboo' with a diaper over her face picked up a bowl to continue the game when the diaper was removed from the play area (Huttenlocher 1974). Mandler points out that this form of conceptual blending is extensively demonstrated in adults (e.g. Coulson 2000).

Despite the claim that image schemas are representations of objects, Mandler acknowledges a 'bias to attend to how things move in relation to each other' that emphasizes 'the "roles" that objects take in events' (2004, 89). Understanding what objects are requires infants to 'know what they do or what is done to them' (89). This is the basis upon which Mandler distinguishes conceptual knowledge of objects (knowledge of what they can do or what is done to them) from perceptual knowledge of objects, which is limited to object recognition.

In a later paper entitled 'Actions Organize the Infant's World', Mandler (2006) again argued for image schemas but this time emphasized that the schemas are action based. Talking about how infants conceptualize during their first year of life, she notes that even with inanimate objects, though 'objects are being conceptualized ... the conceptualization has to do with actions rather than the way we usually think of objects, as having a particular shape, color, size, parts, and features. It appears to be actions – kinds of movements between interacting objects – that are crucial in getting concepts about objects off the ground' (2006, 112). Image schemas, she says, 'are analog spatial descriptions, usually dynamic in character that express primitive or fundamental meanings' (2006, 113). So, for example, one moves along a path; one puts things into containers; one makes something go up or down; etc. All the image schemas are about doing something, or actions. When these image schemas later become explicit in language they show up as verbs or prepositions (depending on the language).

Among the more prominent theorists of cognitive development is Susan Carey, who identifies concepts as 'mental symbols, the units of thought', or more succinctly as 'mental representations' (2011, 113). Carey's dual-factor theory of concepts, outlined in *The Origin of Concepts* (Carey 2009), specifies that concepts are partly but not exhaustively specified by the entities to which they refer. Carey views core cognition, the prelinguistic conceptual repertoire, as rich and powerful. Core concepts are accessible, represented in working memory, and supportive of mechanisms enabling action (i.e. decision and control mechanisms). Her core concepts are not reducible to either sensory or spatio-temporal description; nevertheless one can extract from her account a primary focus on objects, even in accounting for action. For example, in explaining an episode of an infant reaching for crackers in one of two bowls, the recognition of the crackers as objects is noted first: 'Infants make a working memory model of the crackers in each of two buckets', which is accessed in order to make quantitative computations that determine the action of crawling toward one of the bowls (2011, 114). This is a means by which objects are represented as relative to the goals of agents. Carey (2009) nonetheless enshrines movement and action as fundamental in many of the infants' concepts. Motions and interactions among objects – object activities – are taken up in relation to questions concerning attributions of agency and causality. In a chapter on core object cognition she points out that a child's 'attention is allocated to individual objects that are traced through time and space' (70). In fact, she says that research on object files and object-based attention shows that two of the three salient psychophysical signatures are the 'privileging spatio-temporal information over property/kind information …' and 'the capacity to track individual objects through occlusion' (71). When discussing an agent's actions she does so in terms of goals, information, and attention. She claims that 'representations of goals and attentional states are aspects of intentional attributions' (158). Further, intentionality is characterized in terms of *aboutness*, since the agent's intentional state is about the world. Intending and perceiving on this view are intentional states. More importantly here, referring (to indicate objects in the world) 'is another paradigmatic intentional relation' (159).

Spatio-temporal information is the basis of the infant's concept of inert or inanimate mechanical causation. In addition infants have a core concept of agent causality. She argues that infants have 'representations of agents, their goals, their attentional and perceptual states, and their information processing activities' (2009, 159). She also claims that '[a]gents are identified through analyses of their actions' (187). Conception of agency (our own or others') impacts on even core conceptual content: 'young infants represent objects relative to the goals of agents, and infants' representations of physical causality are constrained by their conceptualization of the participants in a given interaction (as agents capable of self-generated motion or as inert objects)' (2011, 114).

Even in the context of discussing agency, however, Carey does not unqualifiedly endorse the importance of activities. She raises the question in chapter 5 (Carey 2009) of whether static features of entities, rather than motion, might also play even a more important role in the assignment of agency and intention, giving the use of eyes and hands as examples of the kinds of features that might be responsible.

We do not wish to critically assess the strengths and weaknesses of Mandler's or Carey's approaches. We only wish to draw attention to the importance of action and causation in their work: the action of the infants in relating to the world, and the action of the world on infants as various aspects of the environment are understood by the child during development. In this way we think Piaget was right minded in stressing that sensorimotor coordination is basic to development. We would add it is not just the sensory systems that need such coordination, but the cognitive ones also. The developing organism must learn to interact with the world in appropriate and reliable ways. This is a basic theme that needs to be articulated systematically and in more detail in work on child development.

ALTERNATIVES OLD AND NEW

The emphasis on action and activity emerging in prominent theories of conceptual foundations may now be compared with what we earlier identified as the situated, distributed, and embodied perspectives, which are increasingly accepted as mainstream cognitive science.

Despite differences in core concepts, emphasis, and method, the situated, distributed, and extended mind approaches all represent efforts to reconceptualize systems, whether representational or not, in ways that show how cultural and cognitive processes are intertwined and such that cognitive activities are made possible or constrained by the specific ways of acting in the learning environment, including social configurations (who is present, the traditions and standards upheld, etc.) and technologies employed (Clark 2003; Greeno 1998; Hutchins 1995a, 1995b). The centrality of activity to these accounts is summarized well, as follows: 'A theory of situated cognition suggests that activity and perception are importantly and epistemologically prior – at a non-conceptual level – to conceptualization. An epistemology that begins with activity and perception, which are first and foremost embedded in the world, may simply bypass the classical problem of reference' (Brown et al. 1989, 41). Such approaches attempt to circumvent the problem of reference by showing how cognition and the world are mutually interconnected in complex causal relations. So, for example, Brown et al. (1989) suggest that cognitive 'representations are … indexicalized in the way that language is … they are dependent on context' (1989, 36).

Situated-cognitive perspectives overlap in some important ways with embodied-cognitive approaches, and they are frequently discussed together. However, situatedness and embodiment are subtly different notions. As Brooks notes: 'An airline reservation system is situated but it is not embodied' (Brooks 1991, 1227). Well-known proponents of the embodied approach to development define embodiment and discuss its significance as follows:

To say that cognition is embodied means that it arises from bodily interactions with the world. From this point of view, cognition depends on the kinds of experiences that come from having a body with particular perceptual and motor capabilities that are inseparably linked and that together form the matrix within which reasoning, memory, emotion, language, and all other aspects of mental life are meshed [p. 1] …This situates cognition within the same continuous, time-based, and nonlinear processes as those involved in bodily movement, and in the large-scale processes in the nervous system. (Thelen et al. 2001, 2)

Both embodied cognition and situated/distributed/extended cognition find historical precedents that help to inform the question of their applicability to cognitive development. That is, the 'new' gang of embodied, situated, and distributed cognitive science is not so new or radical. There is continuity between the means by which cognitive processes are construed in the new distributed and situated approaches and ideas that are central to what was heralded as an entirely new school of thought one hundred years ago, classical American pragmatism. In many cases there are explicit efforts to ground new developments in pragmatist theory (e.g. Lave 1988). More commonly, pragmatists (usually Dewey) are identified along with representatives of other theoretical perspectives (Vygotsky 1978; Gibson

1979) as providing inspiration for the view of knowing as process or activity rather than as something acquired and stored, and as implicating 'the situation' in all activity. The centrality of activity to the perspectives from which situated and embodied perspectives draw is obvious in the term 'activity theory' (Leont'ev [1959] 1981; Vygotsky 1978), and in 'persons acting' as unit of analysis in Lave's *Cognition in Practice* (1988).

Although *pragmatism* is cited frequently as the framework of inspiration for more contemporary cognitive approaches, and although all pragmatists emphasized the *active nature* of cognition to some extent, Dewey is the clearest and most consistent advocate of the centrality of action. Dewey also provides the most coherent account of 'situation', and his description enables us to identify important differences from contemporary theories of cognitive development that acknowledge the importance of action without *prioritizing* it.

What is designated by the word 'situation' is not a single object or set of objects and events. For we never experience nor form judgments about objects and events in isolation but only in connection with a contextual whole … In actual experience there is never any such isolated singular object or event; an object or event is always a special part, a phase, or aspect, of an environing experienced world – a situation. (Dewey 1938, 66–67)

Earlier, in critiquing the reflex arc concept, Dewey describes acts as coordinated events. The 'defect' in traditional psychology's theoretical treatment of the reflex is the assumption that 'sensory stimulus and motor response [have] distinct psychical existences, while in reality they are always inside a coördination and have their significance purely from the part played in maintaining or reconstituting the coördination' (Dewey 1896, 360). Thus 'objects' have meaning (reference) only as aspects of situations and actions are coordinations within contexts. Further, part of the situation or context of objects is established by what the objects do or can do. The objects in situations include other experiencing organisms – other agents – exhibiting coordinated sequences of action toward goals (problem resolution). The emphasis on contextual wholes and coordination does not mean that experience is an undifferentiated morass. Distinctions between objects and kinds of objects and relations between objects are made, but always with regards to specific purposes, from a point of view, and in keeping with previous identifications and interpretations relevant to that context (Thayer 1968). That is, the basis of distinction is functional rather than structural: there is a transaction between the perceiver's goals and the world in which she acts that is coextensive with experience itself.

GIBSON AND ACTIVE PERCEPTION

The interrelations between perceivers and the world were highlighted in J. J. Gibson's *The Senses Considered as Perceptual Systems* (1966). Gibson stressed that perceivers had to be active, and that active pick-up of information from the world took place as perceivers became attuned to the properties of optic arrays that specified information about the world. In later work (1979) Gibson stressed that the information taken up by the perceiver from the world included those properties in the environment that afforded the perceiver a chance to act in some appropriate way. Susan Hurley (2001) characterized Gibson's later ecological approach as active only in an instrumental sense. That is, perceivers were active only in that they sought perceptual information from the world in an active manner. But contrary to this view, Gibson's idea was that perception was or became attuned to environmental affordances. He actually highlighted the idea that what was important to the perceiver were the aspects of the world that afforded them opportunities to act in appropriate ways. It could be thought that in this way he turned all perceivers into agents.

Reference (or a perceiver's interaction or attachment to the environment) for Gibson was determined by the flow of information from an environmental optic array to a perceiver who picked up the information over time in such manner that it could be used in subsequent action that afforded the perceiver some specific way(s) of acting. That is, the perceiver had the information so that she might do something or act with regard to the environmental aspects it specified. Objects were referred to in perception, only in terms of the higher-order properties that they afforded a perceiver for acting. The information that remained constant (invariant) during this process was elucidated in terms of the causal specificity that linked an environmental optic array, a perceiver's action, and environmental properties relevant to the perceiver's acting in specific ways. Notoriously, Gibson himself decried the usefulness of talking about or including any cognitive or conceptual processing. He dismissed processing (at least in part) because he thought it had to be empiricist and based on sensations, and so made everything else constructive. We have seen above that the dismissal of sensory empiricism is agreed upon by many. In so far as he thought that we didn't need to take account of anything cognitive, his position is problematic. Minimally, one needs an account of perceptual attention. We also believe that one needs an account of cognitive or conceptual functioning, even if *representation* and/or *concept* may be the wrong theoretical tools to do this work (Ramsey 2007; Machery 2009).

The Gibsonian line on active perceiving has been taken up by many later thinkers. Recently it has been expanded and generalized in Andy Clark's *Supersizing the Mind* (Clark 2008). Clark stresses both embodied cognition and organism–environment interactions: 'The embodied agent is empowered to use active sensing and perceptual coupling in ways that simplify neural problem solving by making the most of environmental opportunities and information freely available in the optic array' (Clark 2008, 17). Clark suggests that the best model for this active, embodied agent is 'continuous reciprocal causation' between the embodied cognizer and the environment in which she is situated (24). This model of working causes is basically a loop mechanism coupling the agent to the environment. He stresses that often times it is important to treat this agent–environment reciprocity as one mechanism (129–131). Ultimately, for Clark, meaning for an organism is tied to action. The 'grasp of meaning [for an organism is] the range of ways an encoding … poises us to act (where acts include but are not exhausted by acts of thinking and inferring) in the world' (56). Clark sees no problem in tying a Gibson-like program to theorizing about the neural or computational mechanisms of cognition.

COGNITIVE GRAMMARS

In discussing Mandler above, we noted she relied heavily on the work of Lakoff, Johnson, Núñez, and others for her central ideas about infants' image schemas or concepts. Much of the work of these thinkers deals with language in various domains, e.g. metaphor. Mandler, of course, since she deals with preverbal infants is not tying her image schemas to language in any way, though she suggests, as we noted, that language use grows out of non-linguistic schemas. But it is of note that most of psycholinguistics is concerned with reference to objects, and the contribution that objects make to the content of linguistic representations. Like the logical tradition of old, many linguists find it difficult to think of referring (or inferring) as an activity that takes other activities as its source or goal. Their examples and discourse often assume that the paradigmatic case of reference is naming or picking out objects. Even cognitive-grammar theorists find it difficult to break the old habits of thought. For example, Ronald Langacker (2008) wants to represent verbs as temporal processes diagramed by a time arrow tying together static sequential states, where the set of states are the basic objects for cognition (110 and 152–153). This is like saying activities are nothing but (probabilistic) state transitions, where the states are the focal entities and conceived as static. (For a critique of this way of thinking, see Machamer 2004.)

In linguistic terms, we want to stress that what we need is a theory of how verbs, representing activities, are often more important than nouns for the meanings of sentences, or, more generally, how they are important for meaningful linguistic acts. Michael Tomasello's 'verb island' constructions are examples of how verbs might be shown to take precedence. 'The verb island hypothesis proposes that children's early linguistic competence is comprised totally of an inventory of linguistic constructions of this type: specific verbs with slots for participants whose roles are symbolically marked on an individual basis' (Tomasello 1999, 139).

Buresh et al. (2006) claim that the importance of verbs lies in their function of describing relations. This stress on relations may seem to take us back to the logical models, which were thought to provide the semantics for verbs or functions in terms of relations among objects. But though these developmental theorists talk the old language, it is clear that their intentions are to include actions as important if not central. Their claim is that among the more important relations described by verbs are the actions of other people, especially those that describe agents acting on objects. They review evidence suggesting that infants by the age of six months represent 'grasping as an object related activity'. A more complex form of representation involves acts of attention, indicated, for example, by pointing or gazing. An infant's orientation to changes of gaze is taken frequently as 'evidence of understanding the act of looking' (211). Woodward (2003) tested orienting responses in seven-, nine-, and twelve-month-olds by having a woman gaze at toys, then reversing their position and gazing again. The finding was that infants at all three ages tracked the shifts in gaze, though only the older infants gazed longer at the objects gazed at. After twelve months, infants use their knowledge of action–object relations in interpreting both emotions and language (Buresh et al. 2006). But it is the relational structure of activities not being tied to specific objects that accounts for their flexibility and foundational status. The authors note:

Infants do not seem to begin with the general expectation that all human motions will be object directed. Instead, they seem to discover the nature of particular actions, beginning with familiar actions, such as grasping and gazing. These particular actions are so ubiquitous that they have become, for adults, metaphors for more abstract intentional relations (e.g., 'The prize was just beyond my grasp', or

'I see what you mean'). However, adults are not limited to understanding certain canonical actions as relational. Instead, we can interpret the same scene through different lenses, focusing on the overt motion or the underlying relations, considering descriptions at different levels, or from different perspectives. (Buresh et al. 2006, 212)

The conclusion from several years of research is that during the first year of life infants 'begin to analyze the relational structure of human actions' (214). This analysis is expressed in their earliest verb use; however, the research also suggests that children's ability to represent action precedes their use of verbs by up to a year. Their view is that 'prelinguistic action knowledge sets the conditions for learning words because it enables the child to represent the relational structure of events' (222).

CONCLUSION: ACTIVITY, REFERENCE, AND REALISM

We suggest rephrasing claims about the relational structure of verbs or events by saying that often we attend to activities or actions without much or any attention to what objects or entities are involved. The use of *event talk* and treating verbs as relations are ways theorists have tried to turn talk of activities into talk about objects. Events were just space–time-extended objects.[2] Verbs merely functioned to point out relations among objects. However, object indifference and attention to activities are really complex subjects. We may find some purchase on the idea that activities are centrally important by looking at some common-sense examples. If one wishes to learn good form for golf, one needs to pay attention to a model or demonstration of good form, good actions. This of course presupposes that one can recognize good golf-game form. Or, take a horse race. If we make a bet, we might want to pick out and track the specific horse on which we bet. But we are actually only interested in how fast the horse is running relative to the fastness of the other horses on the track. Our focus is on relative positions and changing relations within the group. This requires individuated horses, but attention that is directed principally to differential motions and relative positions.

To briefly return to the evidence from studies on mirror neurons, the fact that mirror neurons fire when a specific action is performed by oneself or another, suggests that most often it does not matter who performs the action. It does matter what the action is and what objects are involved in the act. Studies demonstrating enhanced effects in experts also suggest

---

[2] For a discussion of events and options for thinking about them, see Casati and Varzi 2010.

that the person's learning history for the activity also matters (see Brass and Heyes 2005, for review).

Our main point here is that developmental theorists, linguists, and psycholinguists need to think more about activities and how these connect relations and the various roles that objects may play in such relations – even whether talking about relations in all cases is what we should be describing when trying to explain cognition, reference, and word use.

When we examined Gibson's position, we noted that reference could be construed as a causal relation between perception of affordances and actions. Correctness of reference (the truth surrogate) was tested by the appropriateness of actions. Appropriate actions are those that are coordinated with features of the environment, and that allow an organism to achieve its goals or purposes in reasonably reliable ways. Reference is secured when an organism learns to interact reliably and effectively with its environments. Reliability has nothing to do with truth. It is about coordinating perceiving, thinking, and acting. For such coordination, as for life in general, muddling through (or satisficing) is enough. This is in line with Raftopoulos' (2009) take on perceptual systems, when he writes that our perceptual system needs 'constraints only to ensure that perception is coordinated with action; they aim to provide a reliable guide for on-line action in interactions with the environment. When it comes to successful action, satisficing and not truth is what matters' (327).

It should be emphasized again that we are not arguing on behalf of any of the theories we have discussed. Ours is not a critical assessment, though along the way we have suggested making changes to theories so that they might better accommodate action and activities. What we stressed is that theories of perception, cognition, and language ought to be theories about activities and actions. This we believe is right, regardless of the details. Our belief is that the major focus of human cognition should not be on what entities are, but what they do – how they act and interact. For the most, and for the most important, part, humans do not care about identifying objects as regarding what kinds of entities they are (object recognition), but they care about what such objects do and how they affect us and enter into causal relations with other objects. There, of course, are specialized tasks that make object recognition primary. Recognizing some person at the airport when you don't know them and are picking that one up makes recognition crucial – in many other cases too. However, most of the time we are interested in what the actions of objects are or their activities. As the pragmatists might have said, doing is more basic than being.

Or differently, we are not thrust into the world as being (as an object), we are born into a world where we must act (or die).

It is this interaction with the world that supports the claim that it is activity that is basic for realism (in some sense). Activities, for the most part, are non-representational and have no conceptual content. They are forms of *knowing how* or procedural knowledge. If it turns out, as presumably it must, that the interactivities of organisms and their environments are largely and for the most part satisfactory, i.e. appropriate for the task or goal at hand, then the actions have to be coordinated with 'real' properties in or activities of the environment.[3] Realism is about reasonably reliable and adequate or appropriate ways of an organism's interacting with its environment. This is some form of realism. Of course, it does not imply that all species of organisms interact with the environment in the same way, or that any one species is epistemically privileged.

To keep pace with new developments, and in line with common-sense reflections, it is time to rethink seriously how childhood cognition, adult human cognition, perception, reference work. These are all activities (cognizing, perceiving, referring) that organisms do. This is the major point of this essay: Stop thinking about objects as primary! The major focus of human cognition is not what things are, but what they do – how they act and interact – and how they impact on human action. We might even say, in slogan form, all knowing is *knowing how* (of which *knowing that* is a special case)!

REFERENCES

Anderson, M. (2003) Embodied cognition: A field guide. *Artificial Intelligence* 149: 91–130.
Barsalou, L. (1999) Perceptual symbol systems. *Behavioral and Brain Sciences* 22, no. 4: 557–660.
Brass, M. and Heyes, C. (2005) Imitation: Is cognitive neuroscience solving the correspondence problem? *Trends in Cognitive Sciences* 9: 489–495.
Brooks, R. (1991) New approaches to robotics. *Science* 253: 1227–1232.
Brown, J. S., Collins, A., and Duguid, P. (1989) Situated cognition and the culture of learning. *Educational Researcher* 18: 32–42.
Buresh, J. S., Woodward, A., and Brune, C. W. (2006) The roots of verbs in prelinguistic action knowledge. In K. Hirsh-Pasek and R. M. Golinkoff (eds.) *Action Meets World: How Children Learn Verbs*. Oxford University Press, pp. 208–227.

---

[3] This argument for a restricted form of realism takes off from the argument in Raftopoulos 2009, 342.

Carey, S. (2009) *The Origin of Concepts.* Oxford University Press.

(2011) Precis of *The Origin of Concepts. Behavioral and Brain Sciences.* 34: 113–124.

Casati, Roberto and Varzi, Achille (2010) Events. In Edward N. Zalta (ed.) *Stanford Enclopedia of Philosophy,* spring edn. Stanford, CA: Metaphysics Research Lab. http://plato.stanford.edu/archives/spr2010/entries/events/

Chandrasekharan, S. and Osbeck, L. (2010) Rethinking situatedness: Environmental structure in the time of common code. *Theory and Psychology* 20, no. 2: 171–207.

Chandrasekharan, S. and Stewart, T. (2007). The origin of epistemic structures and proto-representations. *Adaptive Behavior* 15, no. 3: 329–353.

Clark, A. (2003) *Natural-Born Cyborgs: Minds, Technologies, and the Future of Human Intelligence.* Oxford University Press.

(2008) *Supersizing the Mind: Embodiment, Action, and Cognitive Extension.* Oxford University Press.

Costall, A. (1995) Socializing affordances. *Theory and Psychology* 5, no. 4: 467–481.

Coulson, S. (2000) *Semantic Leaps: Frame Shifting and Conceptual Blending in Meaning Construction.* New York: Cambridge University Press.

Decety, J. (2001) Is there such a thing as functional equivalence between imagined, observed, and executed actions? In A. N. Melzoff and W. Prinz (eds.) *The Imitative Mind: Development, Evolution, and Brain Bases.* Cambridge University Press.

Dewey, J. (1896) The reflex arc concept in psychology. *Psychological Review* 3: 357–370.

(1938) *The Logic of Inquiry.* New York: Henry Holt.

Fadiga, L., Fogassi, L., Gallese, V., and Rizzolatti, G. (2000) Visuomotor neurons: Ambiguity of the discharge or 'motor' perception? *International Journal of Psychophysiology* 35: 165–177.

Gallese, V., Ferrari, P., Kohler, E., and Fogassi, L. (2002) The eyes, the hand, and the mind: Behavioral and neurophysiological aspects of social cognition. In Marc Bekoff, Colin Allen, Gordon M. Burghardt (eds.) *The Cognitive Animal: Empirical and Theoretical Perspectives on Animal Cognition.* Cambridge, MA: MIT Press.

Gibson, J. J. (1966) *The Senses Considered as Perceptual Systems.* Boston: Houghton Mifflin.

(1979) *The Ecological Approach to Visual Perception.* Boston: Houghton Mifflin.

Greeno, J. G. (1998) The situativity of knowing, learning, and research. *American Psychologist* 53: 5–26.

Holt, L. E. and Beilock, S. L. (2006) Expertise and its embodiment: Examining the impact of sensorimotor skill expertise on the representation of action-related text. *Psychonomic Bulletin & Review* 13, 694–701.

Hurley, Susan (2001) Perception and action: Alternative views. *Synthese* 129: 3–40.

Hutchins, E. (1995a) *Cognition in the Wild.* Cambridge, MA: MIT Press.

(1995b) How a cockpit remembers its speeds. *Cognitive Science* 19: 265–288.

Huttenlocher, J. (1974) The origins of language comprehension. In R. Solso (ed.) *Theories in Cognitive Psychology*. New York: Erlbaum, pp. 331–368.

James, W. (1890) *The Principles of Psychology*, vols. 1–2. New York: Henry Holt.

Lakoff, G. (1990) *Women, Fire and Dangerous Things*. University of Chicago Press.

Lakoff, G. and Johnson, M. (1980) *Metaphors We Live By*. University of Chicago Press.

(1999). *Philosophy in the Flesh*. New York: Basic Books.

Langacker, Ronald W. (2008) *Cognitve Grammar: A Basic Introduction*. Oxford University Press.

Lave, J. (1988) *Cognition in Practice*. Cambridge University Press.

Leont'ev, A. N. ([1959] 1981) *Problems of the Development of the Mind*. Moscow: Progress Publishers.

Machamer, Peter (2004) Activities and causation: The metaphysics and epistemology of mechanisms. *International Studies in the Philosophy of Science* 187, no. 1: 27–39.

Machery, E. (2009) *Doing without Concepts*. Oxford University Press.

Mandler, Jean M. (2004) *The Foundations of Mind*. Oxford University Press.

(2006) Actions organize the infant's world. In K. Hirsh-Pasek and R. M. Golinkoff (eds.) *Action Meets Word: How Children Learn Verbs*. Oxford University Press, pp. 111–134.

Palmer, Stephen E. (1999) *Vision Science: Photons to Phenomenology*. Cambridge, MA: MIT Press.

Piaget, J. ([1937] 1954) *The Construction of Reality in the Child*. New York: Basic Books. Originally published as *La construction du réel chez l'enfant* in 1937.

Prinz, W. (1992) Why don't we perceive our brain states? *European Journal of Cognitive Psychology* 4: 1–20.

(2005) An ideomotor approach to imitation. In S. Hurley and N. Chater (eds.) *Perspectives on Imitation: From Neuroscience to Social Science*. 2 vols. Cambridge, MA: MIT Press, vol. 1, pp. 141–156.

Raftopoulos, Athanassios (2009) *Cognition and Perception: How Do Psychology and Neural Science Inform Philosophy?* Cambridge, MA: MIT Press.

Ramsey, William M. (2007) *Representation Reconsidered*. Cambridge University Press.

Rizzolatti, Giacomo, Fogassi, Leonardo, and Gallese, Vittorio (2006) Mirrors in the mind. *Scientific American* 295, no. 5: 54–61.

Suchman, L. (2007) *Human–Machine Reconfigurations: Plans and Situated Actions*, 2nd edn. Cambridge University Press.

Thayer, H. S. (1968) *Meaning and Action: A Critical History of Pragmatism*. New York: Bobbs-Merrill.

Thelen, E., Schöner, G., Scheier, C., and Smith, L. B. (2001) The dynamics of embodiment: A field theory of infant perseverative reaching. *Behavioral and Brain Sciences* 24: 1–36.

Tomasello, Michael (1999) *The Cultural Origins of Human Cognition*. Cambridge, MA: Harvard University Press.

Vygotsky, L. S. (1978) *Mind in Society*. Cambridge, MA: Harvard University Press.

Wilson, N. and Gibbs, R. (2007). Real and imagined body movement primes metaphor comprehension. *Cognitive Science* 31: 721–731.

Woodward, A. L. (2003) Infants' developing understanding of the link between looker and object. *Developmental Science* 6: 297–311.

CHAPTER 8

Personal and semantic reference

Gerald Vision

[M]any of the most Learned and Wise adhere to the New Scheme of expressing themselves by *Things*, which hath only this Inconvenience attending it, that if a Man's Business be very great, and of various kinds, he must be obliged in Proportion to carry a greater bundle of *Things* upon his Back, unless he can afford one or two strong Servants to attend him. I have often beheld two of those Sages almost sinking under the Weight of their Packs, like Pedlars among us; who, when they met in the Streets, would lay down their Loads, open their Sacks, and hold Conversation for an Hour together; then put up their Implements, help each other to resume their Burthens, and take their Leave.

Jonathan Swift on the Academy of Lagoda, *Gulliver's Travels*, Part
III, chapter 6

The good news is that referential theories have progressed to a point at which the literature contains a wealth of subtle and finely tuned data and doctrines appropriately grounded in it. The bad news is that many of those at work on the topic have been laboring at cross-purposes. Despite these important differences, theorists have regularly attempted to forge a unified account, the result of which has typically been to demote one or another of the competing conceptions or, in a case of interest here, to dispatch one of them to someone else's agenda. I shall concentrate on one of those divisions, that between personal and semantic reference (p-reference and s-reference, respectively). Very basically, the difference is that on p-reference the subject (i.e. referrer) is a person or speaker; on s-reference the subject is a linguistic expression. But that description greatly oversimplifies their differences. The distinction opens up a number of heated issues and divisions, including those over the status of definite descriptions, truth-conditional semantic theories, semantics versus pragmatics,

My thanks to the University of London's Institute of Philosophy for providing the premises and resources for writing this chapter.

speech-act theories of reference, and, finally, the relation between thought contents and linguistic contents. I do not contend that there is a clear victor in these struggles, although I eventually offer tentative steps in the direction of a strong version of p-reference. It is based on what I take to be some generally overlooked virtues of p-reference and on the fact that its critics haven't made out their case. Of course, we shouldn't even begin to evaluate the respective positions until we have a more detailed explanation of each of them and have clarified their important differences. But before attending to that, and to make manageable a condensed discussion of a sprawling topic, preliminary remarks are in order.

First, our discussion is confined to singular terms, including proper names, pronouns, definite descriptions (perhaps some indefinite ones), sobriquets, common count nouns such as 'table', 'horse', 'snowflake', mass nouns such as 'snow' and 'gold', and some common non-count nouns such as 'red' (or 'redness'). Typically they will be terms found in the subject position of sentences, but they may be objects in predicate clauses (e.g. 'Charlie' in 'The short straw was drawn by Charlie'). They may also occur in compounds lacking subject–predicate structure (e.g. 'If *Daphne* forgets the sandwiches, the picnic will be ruined'). For convenience I will concentrate on proper names and definite descriptions (hereafter just 'descriptions' unless otherwise specified).

Second, the word 'term' in the phrase 'singular term' can mislead. So as not to prejudge some crucial issues, our singulars may also be tokens or vehicles in thought contents, loosely known as *concepts*. It is customarily simpler to describe what is going on in terms of linguistic tokens, and I shall follow that custom; but it is important to note that this discussion covers *both* language use *and*, if it is distinct, thought content. For simplicity I shall often refer to the subject of a p-reference as a *speaker*, though in fact she may be, among other things, a thinker, a writer, or a hand signaler rather than a vocalizer.

Third, we shall have occasion to mention the bearers of truth-value. I use the terms 'proposition' and 'utterance' loosely to designate them. Readers are free to substitute 'statement', 'judgment', 'thought', or whatnot for those terms. And although later it comes into play that truth-value, as distinct from a thin notion of truth conditions, does not occur at the sentential level, for those who disagree much of what I have to say will go through if one substitutes 'sentence' (or sentence plus a speaker, time, and place) where I write 'proposition'.

Fourth, the literature contains a number of terms fitting our target other than *reference*, including *denotation, designation,* and *naming.* Russell, for example, labeled descriptions 'denoting terms' (1905, 1919), and some have proposed resolving the notorious differences between his views and Strawson's by claiming that referring and denoting are simply different concepts. Because it has also been claimed that each concept is regimented or quasi-technical, those parties may have concluded that there is no way to approach their differences other than to compare their strictly *methodological* pros and cons. This is too facile. Both parties (and those within each side engaged in internecine squabbles) are interested in phenomena too similar and overlapping to be satisfied by so innocuous a resolution. By 'reference' here I intend any sort of singular term-to-world relation (or speaker-using-a-term to world, or even quasi-relation; see below) occurring in the production of a larger sentential or thought unit.

In conjunction with the last point I take it that referential theories are at bottom attempts to gain understanding of reference through a systematization of the notion of *talking about* (or *thinking about*). Some may argue that the *talking about* idiom is much too unruly to serve as a hard datum which even a regimented theory could develop. But although *talking about* isn't an end-all for a theory of reference, I take it as an indispensable starting point. If certain cases of talking about are not what our investigators want to understand, it is not easy to see *what* they are after and why it should have been of such intense interest to philosophers. As a strictly historical matter, philosophical concern with reference has been launched by cases in common parlance in which talking about, referring, naming, and the like have landed us in paradoxes and various other unwelcome stews. (For example, consider Frege on substantive identities, or the puzzles from which Russell's theory of descriptions takes off. Of course, their presentations might have been intended merely as expository devices, but those cases do raise issues that make attention to them philosophically worthy.) It is hard to see how the topic could retain our interest if we stray too far from those data.

TWO SUBJECTS OF REFERENCE

The acknowledgement of p-reference or of s-reference is frequently – though not invariably – accompanied by the view that one or another of these is basic. (Call those making this claim p-referentialists and s-referentialists, respectively.) Thus, P. F. Strawson (1971, 8) declares his commitment to the primacy of p-reference: '"Mentioning," or "referring," is not something an expression does; it is something that someone can use an expression to do.' On the other hand, Peter Geach (1962, 8) clearly plants his s-referentialist flag with the remark, 'Personal Reference – i.e. reference corresponding to the verb "refer" as predicated of persons rather than of expressions – is negligible for logic; and I mention it only to get it out of the way.' And by 'logic' he intends any semantic theory conforming to first-order predicate logic. Within p-reference, we might further distinguish between

(i)   *Strong p-reference.* Semantic reference, if it exists at all, is always determined by (at least an implied) personal reference.
(ii) *Intermediate p-reference.* Personal reference determines some, but not all, semantic reference.[1]
(iii) *Compatibilism.* Both kinds of reference exist, with neither subordinate to the other.

Keith Donnellan's distinction between referential and attributive (uses of) descriptions, discussed in greater detail below, could suggest either (ii) or (iii). His referential uses are all p-references. But he notes that some attributive uses may also be counted as referential, albeit in an attenuated sense. If so, they would be s-referential. This would suggest that the two notions were autonomous at least to some degree.

We can devise an s-referentialist position closely paralleling (i). Geach's view will serve; all reference of account is s-reference. An intermediate s-referential view is not likely to affirm that any s-reference *determines* a personal reference; it would more likely hold that both sorts occur and are equally legitimate objects of study. And this is tantamount to (iii), compatibilism, while allowing the sides to continue to differ on the exact significance of the p-references.

But what more precisely is it for a speaker or an expression to refer? Thumbnail sketches will have to suffice here.

I begin with s-reference. A description or name semantically refers to an individual *only if* the individual is correctly described or so named. This is a necessary condition, but is insufficient. For one thing, occurrences in opaque (*de dicto*, intensional) contexts are typically non-referential. Moreover, it has been held by many that what in Donnellan's terms would be called 'attributive' occurrences of descriptions, and even names, are non-referential; the descriptions so occurring are, per Russell, devices of quantification rather than referring expressions. One might also insist that

---

[1] For example, it may determine reference for incomplete descriptions (i.e. where an accurate description applies to more than a single individual). The issue is discussed below. See also e.g. Recanati 2001.

the denotation of a description must satisfy it *uniquely*. But, as we shall see, for typical conversational occurrences this must be qualified by 'in the relevant context'. A similar qualification is in the offing for uses of names that would non-contextually name more than a single individual.

A referring expression refers *semantically* if the referent is determined on the basis of the application of the token alone. Even if in the unexceptional case the intentions of the speaker are somehow involved, the term rather than the intention fixes the referent. In some cases we can expect that this will run counter to the speaker's intentions. Of central importance for those who promote strong s-referentialism is that it may contribute 'to the identity of the proposition expressed by the utterance of the sentence ... of which [it] is a constituent' (Neale 1990, 67). This may be its contribution to the proposition's truth conditions, and some popular semantic theories are built on the sentential contribution to truth conditions. Other satisfaction conditions, for non-truth-evaluable utterances, may then be understood in terms of the basic truth-conditional case. On the model of singular predications, the proposition expressed by the sentence 'Theaetetus sits' is true if and only if the property (/universal/ predicate) of *sitting* applies to what 'Theaetetus' names. If Theaetetus is standing or doesn't exist, the proposition isn't true. Reference is a necessary but insufficient condition for truth. It is thus conceived as a part of the grander project of devising an enlarged semantic theory for sentences in which it occurs. Of course, s-referentialists are concerned with more than a reference's contribution to satisfaction conditions; they also offer an account of the process which fixes the reference – namely, accuracy of ascription. Nevertheless the contribution to truth conditions is an adequacy condition which any further account of these mechanics of reference must satisfy. p-reference, as explained below, is then regarded as a purely pragmatic phenomenon (if not dismissed altogether, as in the quote from Geach, pp. 163–164) once it is declared that its successes need not suit it for that role.

Differing accounts of descriptions highlight these differences.[2] As noted, defenders of a Russellian theory of descriptions customarily regard descriptions as other than referential expressions, or really any kind of *expression*; descriptions break down into quantifiers and predicates upon analysis of

---

[2] Bear in mind that one may be an s-referentialist, or a non-referentialist, about descriptions and still take any of our positions (strong, intermediate, or weak) with respect to other referring terms such as names or pronouns. In fact, some (e.g. Cappelen and Lepore 2005) have a minimalist semantic theory containing specifications of truth conditions, but regard *all* reference as pragmatic, none of it as semantic.

the sentences containing them. Also as noted, Keith Donnellan's (1966) class of referential (uses of) description challenges the Russellian analysis for those cases. s-referentialists such as Neale and compatibilists such as Kripke (1979) regard p-reference as of no interest to semantics, to be accounted for pragmatically by maxims and principles typified by those in Grice's (1989) account of conversational implicature. But there are divisions even within this camp. For example, while it may be standard to regard Russellian descriptions as quantificational devices, Kripke allows s-references for what he labels 'weak' Russellian languages. Indeed, he doesn't even hold that the Russellian analysis is ultimately correct, expressing doubts, as we do below, about incomplete descriptions. He enters these deliberations only because he makes a case that p-reference, at least in Donnellan's version, isn't a semantic notion and doesn't require us to modify in any way our account of s-reference.

Turning our attention to p-reference, certainly the *intention* of the speaker is paramount. Geach suggests it is sufficient. His remark quoted earlier is preceded by an imaginary conversation in which Smith says to his wife, '[t]hat fat old humbug we saw yesterday was made a full professor'. Intending an individual or having one in mind is clearly central to p-reference, but making it sufficient would result in a very thin concept. For one thing, speaker's intentions have a slot even in s-referentialists' characterizations of their favored concept. Thus, Jason Stanley writes that the *semantic* value of a basic sentential constituent 'is determined by *speaker intentions* together with features of the context, in accord with the standing meaning of that lexical term' (2005, 226; my emphasis). These may be the kind of general intentions involved in any language use, say, on a Gricean (1957) theory of non-natural meaning.[3] Moreover, Kripke (1979) requires a *general* speaker intention in the determination of a semantic referent. In addition, Kripke (1980), who argues for the insignificance in these studies of p-reference, needs speaker's intentions whenever there is a name applying to more than one thing – say, I name my pet tortoise 'Darwin' – and a question arises about which referential chain a certain use of that name falls on. (This also indicates that the distinction between personal and semantic reference is orthogonal to the differences between causalist and descriptivist theories of reference.)

---

[3] On which *S* intends that *H* believe the proposition expressed by the sentence uttered, and that *S* intend that *H* recognize this intention and acknowledge the proposition on the basis of that recognition.

There are additional complications. Oftentimes writers use *having something in mind* as the condition replacing or interchangeable with *intending something*. But I can have something in mind without intending to refer to it. The exasperated teacher who exclaims 'the prankster, whoever it is, will not be punished if he or she comes forward' may well have in mind the student she believes did it, but she scrupulously avoids referring to a particular student. She may sincerely deny that she was referring to Dennis even if she had him in mind.

Thus, we know that intending is centrally involved, but it doesn't yield a satisfactory definition of p-reference. Neale suggests a set of conditions for the referential use of descriptions that, when modified for all p-reference, might be thought to yield the desired account.

(a) There is an object b such that S knows that b is uniquely F [or uniquely named N];
(b) It is b that S wishes to communicate something about;
(c) 'The F' [or 'N'] occurs in an extensional context;
(d) There are no pronouns anaphoric on this occurrence of 'the F'. (1990, 65)

The teacher satisfies (a)–(c). She fails (d), but if she had said only, 'the prankster will not be punished' (which removes the anaphor) she would have satisfied the whole set while having Dennis in mind, but without p-referring to him. So, we have not yet obtained a sufficient condition for p-reference. Perhaps what is needed is not an additional condition, but rather the absence of various types of defeaters. However, I shall not attempt to fill out the account. One complication in achieving an account is that p-reference may seem to be a kind of subsentential speech act. We might then compare the requirements of p-reference to the rich sets of conditions one finds in speech-act theory, which would naturally include the condition that the speaker also have the intention that the hearer recognize his intention. However, if we extend personal reference to thought, where no second party is involved, there is no distinct recipient of the personal reference to which that intention is directed.

Rather than pursuing these knotty details, I satisfy myself here with focusing p-reference on speaker's intentions. The issues separating the two sides in this dispute concern only that aspect, allowing us to dispense with a more precise account. None of the disputants with whom I am familiar deny that there is such a thing as what I have been calling *personal reference*. The only question is whether it has a place at the table in the serious philosophical (analytical) study of reference.

To illustrate the differences between p- and s-reference, some advocates of personal reference have highlighted those cases in which it succeeds despite a misdescription or a misnaming. Let's follow the general thread of the literature by drawing examples from descriptive referring expressions. (Whereas it was noted that an s-referentialist about descriptions may or may not be an s-referentialist with respect to other expressions, p-referentialists tend to be so for all expressions. Moreover, once again, Russellians as such will maintain that descriptions aren't referring expressions at all.)

I begin with a familiar example. S says 'Smith's murderer is insane' meaning to pick out Demented Dan, the suspect in the dock. However, it turns out either that Smith died of natural causes, committed suicide, or (we may suppose) Sane Sam did the deed; Demented Dan simply was caught mutilating a corpse he found in the gutter and was wrongly charged with Smith's murder. We may suppose that S is speaking to H and intending to draw H's attention to the erratically behaving person in front of them, who is in fact insane. S can be said to be referring to Demented Dan. There are a number of similar cases. S might have misnamed someone – say Sally, by calling her 'Sandy' – or S might refer to someone facetiously or ironically with a description S and H know doesn't accurately apply to the person (e.g. 'generous Ebenezzer'); or to avoid reprisals from a tyrant S could refer to a universally known pretender as the king. In fact, when a British commander, upon capturing a province on the Indian subcontinent, sent back to the home office a terse cable reading 'Peccavi' (Latin for 'I have sinned'), a case could be made that *he* referred to Sind. Clearly, in such cases (with the possible exception of the last one) it is the intention of the speaker rather than the applicability of the expression that carries a perfectly understandable and commonly acknowledged p-reference where it misnames or misdescribes the referent.

However, this distinctive feature, regularly emphasized to explain p-reference, has also been the key to the s-referentialist's dismissal of it as having none of the semantic import essential to the primary philosophical analysis of reference. The claim is that the proposition made with 'Smith's murderer is insane' is false – Sane Sam is clearly not insane. The p-reference is powerless to undo that result, because it doesn't contribute to the truth conditions of the resulting proposition. p-referentialists have replied by trying to show that there are senses in which an uttered proposition can be true despite a mislabeling of the referent. To counter this, s-referentialists may point out that being *true of* is not the same as being true, *simpliciter.*[4] In fact, I believe that intuitions about when a proposition of this ilk is true are inconclusive. Pondering the case of staying on the safe side of the pretender, I'm less certain that a proposition expressed by the sentence 'The king is in council' is strictly speaking untrue. Or, to take a case even more favorable to p-referentialism: Suppose everyone falsely believes that a certain strangely marked bit of rock is an annelid fossil, and it is never in the whole of history discovered that in fact it is not a fossil of any kind. Would the misdescription of it as 'fossil X' make everything said about it using that misdescription false (e.g. 'if you're looking for the annelid fossil, it's located in the museum's Huxley room')? Nevertheless, it is at least unclear that the p-referentialist has an effective reply. Let us then agree that the p-referentialist hasn't shown that p-references under misdescription contribute to the truth conditions (or any other satisfaction conditions) of the larger utterances in which they occur. This is a basic part of the s-referentialist's case that p-reference is semantically irrelevant, merely a pragmatic phenomenon. I shall call it 'the flagship objection'.

To clinch the case that p-reference is exclusively pragmatic, s-referentialists may describe in detail just how p-references fit the recipe for a Gricean (1989) conversational implicature. This isn't perfectly conclusive. Typical tests for conversational implicature involve inferences we can expect to be made even on the basis of what the s-referentialist would regard as semantic features of propositions containing s-references. Thus, that an expression satisfies a significant portion of these tests may show nothing about the expression being non-semantic. A few examples may help. Grice's (1989, 26–27) Cooperative Principle – 'Make your conversational contribution such as is required, at the stage at which it occurs, by the accepted purpose or direction of the talk exchange in which you are engaged' – will generally be satisfied by the semantic element of one's conversation, as will the maxims of Relation ('Be relevant'), of Quality ('Try to make your contribution one that is true'), of Manner ('Avoid obscurity … Be brief'), and of Quantity ('Do not make your contribution more informative than is required'). It will normally even satisfy subsidiary maxims, such that of sociality ('Be polite' [28]). Satisfying any of these doesn't show that the expression isn't semantic.[5] When fortified by

---

[4] s-referentialists may regard 'true of' as a near synonym of 'predicable of'. But as John Cook Wilson (1926, vol. I, 115–116) has shown, to say that such-and-such is *said or asserted of* something (which holds for *true of* as well) is always conveyed by a propositional 'that' clause (in which the subject may be considered as occurring outside the scope of any potential quantification over the clause). What is *true of* Demented Dan is not insanity, but rather *that he is insane.*

[5] The same may be said of Neale's *justification requirement* (1990, 78) which he appends to the Gricean account.

the flagship objection, it may be helpful to show that p-reference satisfies Grice's recipes for conversational implicature. In light of the fact that so many allusions to the pragmatic are little more than hand-waving, beyond showing that p-reference isn't of particular philosophical interest, providing the details of its pragmatic status strengthens one's case that this is its proper home. But showing that p-references fit the Gricean recipes can't succeed as a stand-alone objection. The s-referentialist's main grounds for dismissing p-reference must be a more direct objection like that of the preceding paragraph. Thus, despite Kripke's (1979) acknowledgement that his reasons are largely methodological, so that it is important to find the right slot for p-reference, both his and Neale's rejections of the semantic significance of p-reference must be grounded in the flagship objection.

P-REFERENCE REDUX

It is important to remind ourselves that, whereas both p- and s-referentialists have emphasized mislabelings – each for their own purpose – normally, and in the larger share of occasions, there is no mislabeling. Nevertheless, the speaker will have the very same particular intentions (to *mean* that individual) that are taken to constitute the p-reference in the abnormal cases. The same speaker intentions which determined the making of a p-reference are also present here, although here the referent of any p-reference will coincide with that of the s-reference. For these cases we can regard them as making both p- and s-references, or we can hold that the presence of the s-reference trumps and thereby eliminates a p-reference. The latter strikes me as difficult row to hoe; all the factors making for a p-reference are still present. Of course, one could add that a necessary condition for a p-reference is that there be no s-reference made to the very same thing. But then why not argue that a necessary condition for an s-reference is that no present p-reference has the same referent? So, let us see how things turn out if there is both a p- and s-reference in the normal cases.

This opens an avenue for the p-referentialist to argue that the p-reference is essential to the normal cases. We still have the cases in which the two modes diverge in their respective referents, the mislabeling cases, but the p-referentialist may now be able to write those off as misfires in the discovery of the true function of p-reference. Mislabelings are useful for displaying how p-reference differs from s-reference, but it might be contended that they do not disclose the true essence of p-reference: they show only that the intentional element of reference is so central that it can overcome its standard referential conduit (i.e. correctly labeled tokens). Of course, no such case has been made out at this point, but its prospect will have been opened by noticing the prevalence of p-references in cases of correct labeling. In what follows I want to inquire further into what that essence might be. The point of the above is only that the line of argument we have just exposed gives a very different explanation of the division between p- and s-reference, and opens the possibility that, in a Strawsonian vein, except for the throwaway abnormal case, the p-reference is the determinant of what we consider an s-reference.

To begin a new line of inquiry, notice a distinction between two approaches taken by referential theorists:

(1) emphasizing *the end product* – the referent; and
(2) emphasizing *the production* – how the reference gets produced.

The distinction is far from absolute and may be fuzzy at the edges, but s-referentialists seem mainly attracted to (1), p-referentialists to (2). Of course, s-referentialists may also claim that they have given an account of how the reference is produced; it results from the accuracy of the description, name, pronoun, etc. But in light of the flagship objection, the ultimate test for *genuine* reference is how the (linguistic or thought) token, the end product, contributes to the overall semantics. And despite the fact that I have concluded that the effort to turn propositions with mislabeled p-references into truths hasn't succeeded, the very fact that p-referentialists make the effort indicates their acknowledgement of an obligation to show how their referents can play a role in truth conditions, and thereby contribute to a sentence's semantics. Each is an attempt to forge the unified account out of the disparate approaches to which I alluded at the outset. But despite these sometimes perfunctory efforts to acknowledge the features emphasized by their opponents, there persists a difference between one or another of these Ür-motivations in the various treatises on the topic.

To illustrate with a specimen of (1), Neale follows Grice in distinguishing between

(i)   the linguistic meaning of an expression ζ; and
(ii) the semantic value of a dated utterance or thought content ζ.

Suppose that ζ is a whole proposition, uttered or thought. We will then want to distinguish further between

(iii) the proposition expressed by speaker S; and
(iv) the proposition meant by speaker S. (See e.g. Grice 1989, 24–26; Neale 1990, 92.)

Notice that (i)–(iii) are products, instances of (1). (iv) may not be an actual product, but it is an intended product. It is on the basis of evaluating these products that Grice and Neale make their case. But let us explore a bit further why it is that the p-referentialist will be unsatisfied with that and turn instead to elements falling under (2).

I can imagine that the p-referentialist's first thought is prompted by the issue that may have motivated the farcical sages at the Academy of Lagoda. It is a feature also prominent in Grice's (1957) earlier ruminations on language. A subclass of physical marks or objects have a special characteristic – they signify or point beyond themselves in a way that is not captured by unadorned causal inference. Walking in the woods I see trees, grass, birds, stones, and so on. But they are nothing else. I can make causal inferences about how they got there, or why they have their distinctive characteristics, or comment on their aesthetic features, but there is no further sense in which they signify something else. However, some marks, like those on this page, have a different life, and can connect in a way to parts of the cosmos, sometimes very distant parts, that are not embedded in their strictly physical embodiments. For the p-referentialist, the words of the s-referentialist, stripped of anything more, fall into this first class: they are physical marks, nothing more. They might as well be footprints left in the sand by shore birds, or detritus washed up by the tide. Thus, even in the case of correct labeling, speaker's intentions will have the role of yielding signifying power to those expressions or marks.

The missing speaker's intention in much of the physical world is not the *general* one mentioned by Kripke. That intention would merely distinguish the meaningful from the meaningless. Rather it is the quite specific intention that directs those expressions and marks to a particular target ('what they mean' in one sense of that phrase). Of course, there will be mislabelings; anything that has a standard use is vulnerable to two non-standard varieties of use – misuse and extended use. I can't think of a single useful item of which that isn't true. For example, the computer on which I am now composing this chapter can be used to hit an intruder over the head or as a doorstop. But that does nothing to disqualify p-reference for its function of making the marks more than generally meaningful. The satire of the Lagodans works because of their ludicrous method for avoiding the mysteries of how brute physical sounds can refer.[6] But the concern about how the basically physical manages to pull off the trick is legitimate and a fit subject for further inquiry. It is precisely why tough-minded physicalists and various shades of other naturalists regard both semantic evaluation (/meaning) and intentional states as prima facie problems for the natural world, and as clear candidates for a naturalistic makeover.

Thus, it is plausible to say that the leading motivation of a p-referentialist is to understand the mechanisms by means of which words and thought constituents can put us in virtual contact with stuff beyond those vehicles. This divides into two points. The first is how my words can signify anything at all, that is, how words gain the power in general of signifying, or how they become meaningful. In this chapter we have started from reference, thereby assuming that this issue has been somehow resolved. The second question, the one that concerns our competing referentialists, is how those tokens can reach the 'right' target. For example, how can my *thought* about Barack Obama (or my *use of his name*) reach Obama rather than George W. Bush, or Churchill, or, for that matter, a rock on the moon? This too is part of the larger problem of how thoughts or words can have significance; but the significance in this case is not *bringing something to mind* (as in traditional notions of *sense* and *meaning*), but sending a 'thought dart' or a 'word dart' to one particular part of the world, a part external to my thought or to my saying or its vehicle. So the main question is how it gets brought off. One way of putting this is that the primary focus of the advocate of p-reference is the engine of reference. Given this concern, the further interactions with other linguistic or cognitive devices, such as truth, while important, are to be handled only at a higher level of theorizing. They are a result to be achieved with p-reference in hand rather than an adequacy condition for undertaking the inquiry in the first place.

But if this much is correct, a next step might be to claim that there is no reference at all without p-reference. Since, as we have noted, the conditions for p-reference are there (and, thus, we may conclude, p-reference itself is there) whenever an s-reference is made, we might consider p-reference as the engine that makes s-reference possible. Without a p-reference, it might be said the words are just marks with no life at all. As Donald Davidson (2001a, 108) put the point, philosophers,

---

[6] Could Swift have had in mind passages such as the following from Locke's *Essay* (1700)? 'Another abuse of words is the setting of them in the place of things, which they do or can by no means signify' (III, x, §17)? Or even the admonition of words standing in the way of clarity by his friend Berkeley (see *A Treatise concerning the Principles of Human Knowledge*, Intro. §§22–24).

psychologists, and linguists need 'abstract' languages for the study of language. However, for the rest of us,

> we all talk so freely about language, or languages, that we tend to forget that there are no such things in the world; there are only people and their various written and acoustical products. This point, obvious in itself, is nevertheless easy to forget, and it has consequences that are not universally recognized.

There is a sense in which this gives the p-referentialist a more direct investment in the examination of reference than that of the s-referentialist. If our study is really of *reference*, then it is good advice to begin without preconceptions about where it must end up, and in particular without preconceptions that it link up with projects from which it is distinguishable in principle (or so one can plausibly argue). If we begin from the *assumption* that reference must serve a purpose in the determination of satisfaction conditions (especially truth conditions) in a more comprehensive semantic theory, we are really studying how reference contributes to a larger project. In general, requiring these sorts of connections to distinguishable theories is not a fruitful precondition to place upon any directed inquiry of a particular phenomenon. To put the point in a way the s-referentialist is likely to dispute (but which I can't see any way around), she is in effect starting from a semantic inquiry into satisfaction conditions, breaking those down into its elements, one of which is baptized the referential element because of its clear overlap with what we ordinarily deem reference, and then seeing how that element fits with the rest. That she then turns to pay close attention to that element in isolation, and denominates her results a theory of reference, bears only indirectly on the study of reference *pure and simple*. I am not claiming that her enterprise is not worthwhile, but only that the p-referentialist has a point in claiming that it is not purely a study of reference. Nor am I claiming that, when the standard and non-standard cases are sorted out, the p-referentialist will be able to show that p-references contribute to the satisfaction conditions of the wholes of which they are constituents. Indeed, the recognition that p-reference seems to occur even when the referent is accurately labeled by its word or thought vehicle, is the start of a way to see how its integration into a semantic theory might work – but it is certainly not a guarantee of success.

WHERE NORMAL P-REFERENCE WORKS BEST

Thus far we have been concerned with outlining a p-referentialist's claim and motivation. But we have yet to examine what support the view might garner. Unfortunately, I must be brief. But a major source of support has been the difficulty of explaining commonplace occurrences of incomplete descriptions (for Kripke 1979: '"improper" definite descriptions') without it. They are definite in form, but their formulations alone do not strictly satisfy Russell's uniqueness condition. For example,

'The table is covered with books.' (Strawson 1971, 14)
The congressman changed his vote.
The window is stuck.
The car won't start.

There is more than one table, congressman, window, and car. So we must be relying on something beyond the words uttered to get us to isolate the right one. Consider Strawson's example in which 'Cx' = 'x is covered with books' and 'Tx' = 'x is table'. Then upon analysis of the whole sentence in which it occurs, Russell's '$(\iota x)\ Tx \dots$' may be defined 'in use' in the first two conjuncts of the open sentence following the initial quantifier below.

$$(\exists x)\{[Tx \ \& \ (\forall y)(Ty \supset (y = x)] \ \& \ Cx\}.$$

But we may wonder, wherein lies the condition that if y is T, then $y = x$, the uniqueness clause? Whether the description refers is beside the point, truth conditions are not: the orthodox Russellian is not excused by the claim that descriptions are devices of quantification rather than referring expressions. Something must account for the analysis' uniqueness clause even for orthodox Russellians. The p-referentialist has a straightforward answer: the speaker has an intention to refer to a particular table, which thereby precludes other tables. What account can the s-referentialist in general, or the Russellian, give?

Neale mentions two s-referentialist options, which he calls *explicit* and *implicit*. On the explicit approach the quantificational description is elliptical, say, for 'the table over there'. On the other hand, the implicit approach finds no verbal complement to flesh out the description's use. Rather, it relies on the context to restrict the domain of quantification. As Neale explains it, '[o]n the implicit approach, the domain of quantification might be restricted to (e.g.) objects in the immediate shared perceptual environment' (1990, 95).

Now, as others have pointed out (e.g. Donnellan 1968), the explicit approach fails. Moreover, it fails for some of the same reasons that Kripke (1980) showed that descriptivist theories of name reference fail; there are too many distinct completing phrases to flesh out the ellipsis. How is the choice between them more than arbitrary? *S* could complete her phrase by 'the table in front us', or 'the table you want me to work at', or 'the table in the dining room', or 'the table in your office', or … Even if it is a table in a room in clear view of the speaker and hearer, there are too many divergent ways to fill out the ellipsis. Is the description 'the table in this room', or 'the table in room 248 of Senate House' or 'the table you're looking at' or …? Each is a different description. Which is the correct one? To suppose that the speaker and hearer are entertaining different definite propositions just because they are likely to complete the ellipsis differently is to be committed to the view that the utterance is ambiguous. But it is implausible to suppose that a typical utterance of any of the sentences displayed above are for that reason intrinsically ambiguous. (Of course, any of them *could* be ambiguous on an occasion. But that would be an exceptional circumstance.) Moreover, it is unclear that this view yields an alternative to speaker intentions; it is plausible that even if a completing phrase could be non-arbitrarily chosen, it would be the speaker's intention that made it non-arbitrary. And then it would be superfluous to go that route rather than just using the speaker's intention to fix the reference of 'the table' in the first place.

For this reason, some may prefer the implicit approach. On it we restrict the domain of quantification. Recall that Neale gives as an example of a restricted domain, 'objects in the immediate shared perceptual environment'. But the example is a clue to the difficulty with this approach. Suppose it is a table in another room on which we are supposed to dine, and this is given as a reason not to do so. Or suppose it picks out a meeting place in the public library at which you suggest we meet. The domain can be restricted in various ways. How does one determine the salience of the table, instrumental to choosing the way the domain is to be restricted, other than through the speaker's intention? Bach's remark that it doesn't matter, 'because the speaker does not intend the description by itself to provide the hearer with the full basis for identifying the referent' (Bach 2004, 221), is beside the point when discussing, at this stage, only the determinants of uniqueness. We are trying to account for the role 'the table' can play in this utterance, and indeed can play in determining the relevant truth conditions. And if the prescription 'restrict the domain of quantification' cannot do the job without the speaker's intention, this strategy too plays into the hands of the p-referentialist.

In light of the failure of explicit and implicit approaches, Lepore (2004) concludes that there is no way to flesh out what is said into a proposition expressed, but that it doesn't matter for the minimalist semantics that he favors. However, the price is that reference is left out of his minimalist semantics. At this stage in our deliberations we can avoid farming out reference by pursuing p-referentialism. Of course, given what has been said, this is no protection against the hearer misunderstanding which table is intended. But a hearer is vulnerable to making mistakes on whatever theory of reference is being proposed. That isn't the liability of any one theory in particular.

I cannot do justice here to the extensive and multifaceted literature on incomplete descriptions that has involved many participants. Suffice it to say that it is considerably more complex than my brief remarks may indicate.[7] A full discussion would need to go well beyond the scope of the present chapter. But I hope to have shown that the p-referentialist has more than a 'picture' of how things might go on its account; she also has some considerable support for her view in this area.

FALLOUT

The adoption of p-referentialism does not resolve all the issues that have grown up around theories of reference. But it is worth a brief review of how the dialectical landscape is changed under the authority of p-reference.

First, asking at the outset whether p-references as a class fit into semantics or pragmatics directs the inquiry onto the wrong path. Our divisions cut across those categories in ways that obscure what is at stake. It is not that we can't make sense of this divide or the classifications they represent, though it is not as neat as its frequently complacent use seems to indicate. But on the view we are tracking p-references would be responsible for any semantic impact reference may have, including the s-references. The class into which the final product falls may depend on whether the mechanisms misfire in certain ways. Conceivably p-references have both semantic and pragmatic products. Looking first for their classification is starting from the wrong end.

Showing how the referent contributes to the truth conditions of its containing utterance would be a welcome result of our inquiry rather than a condition of its undertaking. I haven't attempted to show that, or how, this might be accomplished, though I hope emphasizing that most p-reference is not the result of a mislabeling is a good start. The point to take away from this discussion is that the phenomenon of p-reference by itself doesn't show that it can't be done.

---

[7] See footnote 1 in Lepore (2004) for an abbreviated list of those who have attempted to tackle the issue with respect to its bearing on Russell's theory of descriptions.

Next, consider some different problems that arise concerning the indeterminacy of reference. Of course, there is a general problem, scouted
by Putnam, of applications to ordinary occurrences of the Löwenheim-
Skolem theorem, which would show that any would-be instance of
referring has an infinity of (re-)interpretations (1977). For example, the
reference to the table in our model utterance has an infinity of interpretations within number theory alone. But there is also the less global threat
from Quine's (1960) radical-translation thesis (whose later upshot is ontological relativity) and from Davidson's (2001b) radical-interpretationism,
though the latter officially deems indeterminacy at most a minor problem.
However, notice that all these issues start from an interpretation of *the
product*, the terms used. It is the domain of terms that are susceptible to
multiple translations; this is quite explicit in Putnam. Also a model of reference that compares our use of language with computer states – assuming
that computers have no intentions – is vulnerable to such difficulties. In
fact, this is another instance of a point stressed a few pages ago – sounds
or marks are just noises or patterns, susceptible to the myriad interpretations with which we can enliven them. It is more difficult to see how
these problems are generated if the source of the reference is located in
the speaker's intentions, the producer. Whereas I can always reinterpret
what someone else says, are my own intentions multiply reinterpretable?[8]
Some have argued for radical translation/interpretation even with respect
to the first person (e.g. Hylton 1991), but that is a rather daunting task.
At any rate, the debate takes on a very different complexion than it had
when we thought of the semantic investigator as an informationless cultural anthropologist.

This issue may seem complicated by semantic externalism. I might be
wrong when I say or think 'this water is dirty' if the substance I am demonstrating is XYZ rather than H_2O. However, the problem is the possibility of mistake, not one of an uncontrollable set of reinterpretations
of my intention. Externalism may threaten a certain kind of privileged
knowledge of the content of my intentions, but it doesn't bear on whether
those intentions determine the reference of my terms. Recall that the
prime examples used to illustrate p-reference, and the basis for the flagship objection to it, were precisely cases in which the referring expression
unequivocally referred to something particular, but mislabeled it. And
if a mistake is made, it is the result of a quite definite condition of the

---

[8] Searle (1987) cites this as a consequence of the Quine–Davidson approach. He argues that is a *reduc-
tio* of interpretive semantics.

environment, not a device that tells me that *regardless of what the environment or world is like*, I can always find another, equally good, interpretation of my reference. The latter possibility is more of a threat when we are discussing already-produced vehicles; the individual tokens are reinterpretable only because the domain in which they are members is so. It is more difficult to motivate the predicament when it applies to a speaker's specific intentions.

Third, reference may now be approached as a subsentential speech act. Does this render it pragmatic rather than semantic? The divisions between those categories are not sufficiently well marked to settle that question in the abstract. For one thing, not all semantic theories are truth conditional. For another, various theorists, whose views can be classified as 'use' theories, have argued that meaning, an arch semantic notion, can be identified with, as William Alston (2000) once put it, illocutionary act potential.

Finally, this view brings the mental into reference, and thus semantics, in a major way. Cognizers re-enter the picture from the very beginning. It is not at bottom the medium – words or their counterpart thought vehicles – that matters, but the mindset of the language user or thinker. Linguistic reference is now firmly integrated with thought reference. The usual view is that thought has original intentionality; (natural) language, derivative intentionality. But the point emphasized here is only the central role for reference now accorded cognition. This also changes the polemics over reference. To give one minor example, Kent Bach (2006) seeks to rebut something on the order of the incomplete description challenge, and to restore context, by noting that a speaker's intention can be folded into her general communicative intention, and need not be a separate referential intention. But if the speaker is nothing more than a thinker, one is hard pressed to find a communicative intention (say, by dividing the thinker into communicator *and* audience, or by treating it subjunctively as what the thinker *would have* intended to communicate). In those circumstances none of the maneuvers one might try out is exceptionally promising.

Given this cognitive orientation, here is another tantalizing hint about how the dialectic may have been altered. There has been a long-standing problem about reference to the non-existent, summarized in the dictum, 'One can refer only to what exists' (Searle 1969). There are reasons supporting this: reference is a relation, and relations demand relata. But we regularly suppose that we can think about what doesn't exist – unicorns, El Dorado, Pegasus. Philosophers have introduced various schemes to overcome the difficulty: from the 'referent' side they have devised universes of discourse, departments of language, and quasi-existential statuses such as 'subsistence'; from the vehicle side they have proposed intentional fictional operators, implicit prefaces such as 'Shakespeare wrote that', and monadic reconstruals such as 'referring-to-Xs'. All such proposals seem to me to have been abject failures.[9] My best guess is that a sensible onlooker, uncorrupted by philosophical impulses to conform to a slate of theoretical constraints, would be able to see that these are desperate, ad hoc, maneuvers. How then are we to proceed?

Here is a clue. The worries over the relational character of reference seem, once again, driven by the 'product' notion of reference, that is s-reference. When we turn our attention to 'thinking about' it seems patently clear that the same sort of mental machinery is at work, say, when we talk about a real duke such as Wellington and a fictional one such as Prospero. Shifting the emphasis to the capacities of speakers and thinkers, from that of what speakers produce, may not resolve the difficulties – some may still be bothered by a certain orthodoxy regarding relations – but it seems to relieve some of the pressure. Thinking about what doesn't exist is a natural capacity rational creatures have, and it would seem sensible to take that ability as a starting point of our investigations rather than as a problem that crops up only after we have laid down ground rules designed for adjacent subjects. For starters, we might take our lead from Lewis' suggestion that whereas a relation 'is a property of 'tuples of things ... [T]here is no restriction to actual things' (1999, 10). Still we may not have resolved all the conundrums about *referring to what is not*, as for example, why it doesn't collapse into *referring to nothing* (and thus *into not referring*), but it seems at least that we have shed an albatross when we acknowledge that thought about what exists and what doesn't may be generated by the uniform and non-defective operation of the same type of mental processes.

REFERENCES

Alston, William P. (2000) *Illocutionary Acts and Sentence Meanings.* Ithaca, NY: Cornell University Press.
Bach, Kent (2004) Descriptions: Points of reference. In Marga Reimer and Anne Bezuidenhout (eds.) *Descriptions and Beyond.* Oxford University Press, pp. 189–229.
    (2006) What does it take to refer? In Ernie Lepore and Barry C. Smith (eds.) *Oxford Handbook of Philosophy of Language.* Oxford University Press, pp. 516–554.

[9] For some of the reasons, see Vision (1982), (1985/86), and (1993).

Cappelen, Herman and Lepore, Ernie (2005) A tall tale: In defense of semantic minimalism and speech act pluralism. In Gerhard Preyer and Georg Peter (eds.) *Contextualism in Philosophy*. Oxford University Press, pp. 197–220.

Cook Wilson, John (1926) *Statement and Inference*, vol. 1. Oxford University Press.

Davidson, Donald (2001a) The second person. In *Subjective, Intersubjective, Objective*, vol. III of *Philosophical Essays*. Oxford University Press, pp. 107–122.

(2001b) *Inquiries into Truth and Interpretation*. Oxford University Press.

Donnellan, Keith (1966) Reference and definite descriptions. *Philosophical Review* 75: 281–304.

(1968) Putting Humpty Dumpty back together again. *Philosophical Review* 77: 203–215.

Geach, Peter (1962) *Reference and Generality*, rev. edn. Ithaca, NY: Cornell University Press.

Grice, H. P. (1957) Meaning. *Philosophical Review* 66: 377–388.

(1989) Logic and conversation. In *Studies in the Way of Words*. Cambridge, MA: Harvard University Press, pp. 22–40.

Hylton, Peter (1991) Translation, meaning, and self-knowledge. *Proceedings of the Aristotelian Society* 91: 269–290.

Kripke, Saul (1979) Speaker's reference and semantic reference. In Peter A. French, Theodore E. Uehling, Jr., and Howard K. Wettstein (eds.) *Contemporary Perspectives in the Philosophy of Language*. Minneapolis, MI: University of Minnesota Press, pp. 6–27.

(1980) *Naming and Necessity*. Cambridge, MA: Harvard University Press. Originally appeared in Donald Davidson and Gilbert Harman (eds.) *Semantics of Natural Language*. Dordrecht: Reidel, 1972.

Lepore, Ernie (2004) An abuse of context in semantics: The case of incomplete definite descriptions. In Marga Reimer and Anne Bezuidenhout (eds.) *Descriptions and Beyond*. Oxford University Press, pp. 41–67.

Lewis, David (1999) New work for a theory of universals. In *Papers in Metaphysics and Epistemology*. Cambridge University Press, pp. 8–55. First published in *Australasian Journal of Philosophy* 61 (1983): 343–377.

Neale, Stephen (1990) *Descriptions*. Cambridge, MA: MIT Press.

Putnam, Hilary (1977) Realism and reason. *Proceedings of the American Philosophical Association* 50: 483–498.

Quine, Willard van Orman (1960) *Word and Object*. Cambridge, MA: Harvard University Press.

Recanati, François (2001) What is said. *Synthese* 128: 75–91.

Russell, Bertrand (1905) On denoting. *Mind* 14: 479–493.

(1919) *An Introduction to Mathematical Philosophy*. London: George Allen & Unwin.

Searle, John R. (1969) *Speech Acts*. Cambridge University Press.

(1987) Indeterminacy, empiricism, and the first person. *Journal of Philosophy* 84: 123–146.

Stanley, Jason (2005) Semantics in context. In Gerhard Preyer and Georg Peter (eds.) *Contextualism in Philosophy.* Oxford University Press, pp. 221–253.

Strawson, P. F. (1971) On referring. *Logico-Linguistic Papers.* New York: Methuen. Originally appeared in *Mind*, n.s., 59, no. 235 (1950): 320–344.

Vision, Gerald (1982) A causal account of name reference. *Ratio* 24: 111–130.

(1985/86) Reference and the ghost of Parmenides. *Grazer Philosophische Studien* 25–26: 297–326.

(1993) Fiction and fictionalist reductions. *Pacific Philosophical Quarterly* 74: 150–174.

CHAPTER 9

Reference from a behaviorist point of view

Don Howard

INTRODUCTION: THE SEMANTIC NATURALISM
OF DEWEY AND QUINE

With good reason, the 'linguistic turn' or 'semantic turn' is regarded as one of the enduring achievements of later nineteenth- and early twentieth-century philosophy. Judgments are not to be confused with propositions, and philosophical questions about the relation between language and world are to be disentangled from psychological questions about the workings of the mind and its causal connections to the world. The rise of modern analytic philosophy (and much of later continental philosophy as well) would have been impossible had it not been for the anti-psychologistic revolution of the late nineteenth century, which reshaped the philosophical problem space and reworked the geography of the disciplines; psychology and philosophy becoming separate departments in most universities, the former an empirical, experimental science, the latter not.[1] It is ironic, however, that, at almost the same time, evolutionary naturalism became a prominent current, at least in North American philosophical thought,[2] with thinkers otherwise as diverse in philosophical orientation as John Dewey and Roy Wood Sellars seeking to embed epistemology, especially, in a broadly Darwinian setting.[3] Evolutionary naturalism was somewhat less fashionable in the decades immediately after the Second World War, but by the 1970s it found a new audience, and evolutionary epistemology remains, today, a thriving area of research.[4] In this version of

---

[1] Rorty 1967, Hatfield 1990, 2003, Coffa 1991, Kitcher 1992, Kusch 1995, Friedman 2000, Jacquette 2003, and Heidelberger 2004 all provide helpful perspectives on aspects of this history.
[2] Of course one does not want to forget the central role that evolutionary naturalism played in the work of a continental thinker like Ernst Mach, whose 'biologico-economical' point of view in epistemology owed a large debt to Darwin. Banks 2003 is a helpful recent study.
[3] Dewey 1910, 1925, and Sellars 1922 are representative.
[4] Donald Campbell is the best-known early champion of the newer evolutionary epistemology; see, for example, Campbell 1974. See also the essays collected in Shimony and Nails 1987 for a broader perspective and Bradie 2008 for a briefer overview.

naturalism, biology is the grounding science, but naturalists like Dewey, or
W. V. O. Quine, to name a more recent exemplar (and one who acknowledges a major debt to Dewey), would also emphasize the continuities
between biological and psychological perspectives on knowledge and language, psychology simply picking up the scientific story at that point in
the life of the individual where evolutionary accounts of the species cease
to be useful. For Dewey, that meant functional psychology; for Quine, it
meant the Skinnerian version of behaviorism, or 'operant theory'.[5]

For much of the twentieth century, these two traditions – analytic philosophy and evolutionary naturalism – lived in tension. In the *Tractatus*
(1922), Ludwig Wittgenstein gave canonical expression to the idea that
philosophy and empirical science, evolutionary biology in particular, have,
as a matter of principle, nothing to do with one another. Wittgenstein
wrote:

4.111 Philosophy is not one of the natural sciences.

and

4.1122 Darwin's theory has no more to do with philosophy than any other
       hypothesis in natural science.

Indeed, if philosophy is just conceptual or linguistic analysis, then empirical science is irrelevant to the philosopher's task.

Some thinkers, however, crossed the analytic/naturalistic divide, Quine
being here, too, a prime exemplar. He argued, famously, that his behavioral perspective on language was subversive of the notions of meaning
(sense) and reference that descended from Gottlob Frege (1892) and were
central to analytic philosophy of language.[6] If evidence in the form of verbal behavioral dispositions provides our only access to meaning, then the
assignment of meanings to terms and propositions is underdetermined by
the evidence. Reference is similarly underdetermined by the behavioral
evidence and is, thereby, rendered 'inscrutable'.

One noteworthy response to Quine also crossed the analytic/natural-
istic divide. Philosophers like Hilary Putnam (1970), Saul Kripke (1971,
1972), and Richard Boyd (1979) sought to render reference 'scrutable' and
determinate again by means of a causal theory of reference.[7] The idea is

---

[5] See Dewey 1922 and Quine 1969a, 1973. For helpful background on Dewey's behavioral view of language and meaning see Lee 1973 and Shook 2000.
[6] Quine 1960 and 1968 are the main sources for the well-known Quinean theses of translational indeterminacy and referential inscrutability.
[7] Another and related motivation among proponents of the causal theory of reference was that of
   providing a realist response to the kind of meaning-dependence and meaning-change arguments

that a term acquires a fixed reference, becoming a 'rigid designator', by means of an initial act of 'dubbing' or 'baptism', as when a parent teaches a child the word 'red' by pointing to a red ball – with all speakers' usages of the term being connected to this initial act by a causal process, as when the child tells a friend, the friend tells a neighbor, and so forth.

In what sense these processes are 'causal' is puzzling, likewise in what sense the whole scheme is 'naturalistic'. A more fundamental question concerns how pointing alone can disambiguate reference in the absence of other verbal cues. Am I pointing to the ball's color, its shape, its size, its texture, its patterning? Deeper still is the question how, in principle, a *causal* relationship (read 'physical', 'biological', etc.) can have anything whatsoever to do with the *semantic* relationship of reference. One worries that some sort of category mistake has been committed here.

Reference understood in a purely formal sense, as a mapping from uninterpreted signs in a formal language to elements of a model, is untouched by many of the above-mentioned worries about Quinean referential inscrutability or the limitations of a causal theory of reference.[8] The serious difficulties all accumulate around the notion of reference as applied to terms in natural languages, as these are spoken and used by real human speakers in the real (or even fictional or counterfactual) worlds. Should we be surprised? Alfred Tarski had already cautioned back in the 1930s that the semantics of natural language would be a messier business than the semantics of formal languages, if only because the notion of sentencehood – what counts as a well-formed formula – is not recursively specifiable for natural languages (Tarski 1935). That sentencehood is not recursively specifiable for natural languages reflects the distinctive plasticity of such languages, the fact that they are capable of not just an infinite novelty of form but an infinite novelty of form that cannot be captured even by infinitely open-ended rules. Such a capacity for novelty is seen also in the referring expressions of natural languages.

---

advanced by Thomas Kuhn (1962) and others on behalf of one or another incommensurability thesis. For a discussion, see, for example, Rorty 1976. Yet another motivation – this especially for Kripke – was to fix flaws in Bertrand Russell's descriptive view of reference, according to which referring expressions are to be read as disguised definite descriptions. As Kripke pointed out, we succeed in referring uniquely even if, for example, we know of no unique definite description or when the definite descriptions we employ are false.

[8] But even reference in this purely formal sense is implicated by another of Quine's famous arguments, this for 'ontological relativity' (Quine 1968). According to Quine, ontology – hence reference, too – is relative both to the choice of the 'background' language in terms of which the model is described and, obviously, to the way in which the mapping is constructed.

The non-recursive plasticity of natural language is a point to which we shall return below. It is but one of various ways in which natural language – the language of poetry, song, and everyday human commerce, as well as the language of much of science and philosophy – eludes our efforts to capture language in formulas. Naturalists like Dewey and Quine appreciated this point, which is one reason why both turned from logic to behavioral psychology as affording more helpful scientific tools for understanding how natural language works.

One could be a psychological naturalist about language while taking the elements of language to be mental entities. But Dewey was emphatic on the point that language and meaning are matters of overt behavior, not the speaker's inner mental life. Dewey put it this way in *Experience and Nature* (using the term 'meaning' in a broad way that comprehends both sense and reference):

Meaning is not indeed a psychic existence; it is primarily a property of behavior, and secondarily a property of objects. But the behavior of which it is a quality is a distinctive behavior; cooperative, in that response to another's act involves contemporaneous response to a thing as entering into the other's behavior, and this upon both sides. It is difficult to state the exact physiological mechanism involved. But about the fact there is no doubt. (Dewey 1925, 179)

Quine was very much of the same opinion. In his 1968 John Dewey lectures at Columbia he explained his debt to Dewey this way:

Philosophically I am bound to Dewey by the naturalism that dominated his last three decades. With Dewey I hold that knowledge, mind, and meaning are part of the same world that they have to do with, and that they are to be studied in the same empirical spirit that animates natural science.

When a naturalistic philosopher addresses himself to the philosophy of mind, he is apt to talk of language. Meanings are, first and foremost, meanings of language. Language is a social art which we all acquire on the evidence solely of other people's overt behavior under publicly recognizable circumstances. Meanings, therefore, those very models of mental entities, end up as grist for the behaviorist's mill. Dewey was explicit on the point: 'Meaning ... is not a psychic existence; it is primarily a property of behavior.' (Quine 1968, 185)

Note that Quine, like Dewey, takes the behavioral perspective on language and meaning also to imply the inherently social character of language. This is a point to which we shall return below when discussing Skinner's concept of verbal behavior.

But what does it mean to say that 'meaning ... is primarily a property of behavior'? We do not ordinarily think of Dewey as a behaviorist. He was, famously, a proponent of the functionalist alternative to the structural-ism that dominated American psychology in the late nineteenth and early twentieth centuries,[9] and he was a famous critic of early stimulus–response (S–R) or reflex forms of behaviorism. In his classic paper, 'The Reflex Arc Concept in Psychology', he argued that stimulus and response could not be seen as separate units of analysis, that they had to be regarded from the point of view of their functional relationship as a unitary whole (Dewey 1896). But, in fact, this functionalist conception of the relationship among stimulus, response, and, most importantly, the accumulated results of earlier experience anticipates the later Skinnerian version of behaviorism, known as 'operant theory', for, as we shall see below, Skinner defines the operant as his fundamental unit of analysis in precisely this way, namely, as a three-term functional relationship among stimulus, response, and the history of contingencies of reinforcement.

It is unfortunate that Dewey never developed his behavioral perspective on meaning beyond the kind of programmatic remarks already quoted. The sophisticated notion of 'habit' that is the basic unit of analysis of his psychology (see especially Dewey 1922) would have served as a useful resource for such a task.

Quine explicitly situated himself in the later behaviorist camp of his Harvard colleague, B. F. Skinner, and in works like *Word & Object* (Quine 1960) and *The Roots of Reference* (Quine 1973) he explored in great detail the consequences for semantics of the adoption of a behavioral perspec-tive on language. But even Quine disappoints by the lack of theoretical sophistication in his behaviorism, his being little better than an arm-chair version of Skinnerian operant psychology. Thus, in the well-known argument for the indeterminacy of radical translation in *Word & Object*, behaviorism lives only in the form of Quine's taking as evidential a native informant's verbal dispositions to assent or dissent when confronted by queried uses of native locution, as when the linguist queries 'Gavagai?' on the occasion of his or her noticing a rabbit in the field of view (Quine 1960, 26–79). Worse still, in *The Roots of Reference* and other writings, Quine makes the fundamental mistake of assuming that stimulus con-ditions at receptor surfaces, what psychologists term 'proximal' stimuli,

---

9 The most influential American structuralist was Edward B. Titchener at Cornell; see, for example, Titchener 1896. Functionalism was prominently advocated by Dewey's University of Michigan stu-dent and, later, University of Chicago colleague, James R. Angell; see Angell 1904. Hothersall 2004, 139–174, 361–392, provides a helpful overview of the history.

are what stand in the relevant lawlike relations to the verbal behavioral responses that are mainly of interest (Quine 1973).[10] Only a little sophistication would have been required to understand that more relevant in most situations are what are known as 'distal stimuli', which means in effect, in many settings, ordinary, medium-sized physical objects. If one wants to understand why a subject vocalizes 'chair', in the presence of the chair, then look to the chair and not the photons impinging on the retina. Most surprising of all, in Quine's version of semantic naturalism, is that, in spite of his invoking the authority of Skinner, he nowhere makes use of any of the theoretical apparatus developed in Skinner's own profound and detailed study of language, his 1957 book, *Verbal Behavior* (Skinner 1957).[11]

Like Dewey and Quine, I think that behavior is where meaning is made manifest. Like Quine, I think that a behavioral perspective on semantics is likely to yield results subversive of the notions of meaning and reference that have descended to us from Frege. But I also think that an important opportunity was lost when Quine and other semantic naturalists failed to make use of the sophisticated theoretical apparatus developed in Skinner's *Verbal Behavior*. That the opportunity was lost is, in some ways, not surprising. Even Quine's amateur behaviorism came under swift and strong attack from critics like Jerrold Katz and Jerry Fodor.[12] By the late 1960s and early 1970s, the influence of Skinnerian operant theory was beginning to wane in the face of the cognitive revolution in psychology (see Neisser 1967). And, of course, Skinner's *Verbal Behavior* had, itself, been the target of what is remembered by many – wrongly, in my opinion – as a famously destructive review by Noam Chomsky (1959).[13] Still, this was, I think, an opportunity lost, and my principal aim in the rest of this chapter is to give a sketch of what a more sophisticated operant theory of meaning and reference would look like and to make clear the main ways in which it would constitute a dissent both from the Fregean tradition in semantics and from the other dominant forms of semantic naturalism, such as the causal theory of reference.

---

[10] It would not be wrong to see here, ironically, a vestige of the phenomenalism that dominated right-wing logical empiricism, a view that Quine otherwise famously disputed.

[11] Chapter 3 of *Word & Object*, 'The Ontogenesis of Reference' (Quine 1960, 80–124), begins with a mention of Skinner's notion of operant behavior and cites *Verbal Behavior*, but mainly only for the general notion of the operant and even that is not deployed systematically in what follows.

[12] See Katz and Fodor 1963, Katz 1972, and Fodor 1975.

[13] I am not the only one who thinks that the Chomsky review missed its mark; see, for example, MacCorquodale 1970. After developing some of Skinner's basic technical apparatus below, I will briefly explain what Chomsky got wrong.

AN OUTLINE OF SKINNERIAN OPERANT THEORY

We begin with a primer on the basic ideas of Skinnerian operant theory before looking at how this framework is deployed by Skinner in the study of verbal behavior. Philosophers often learn about behaviorism by reading either Gilbert Ryle (1949) or Chomsky (1959). This is unfortunate, for, like casual readers of Skinner's own more popular work, philosophers thus fail to appreciate fully the nature and extent of the Skinnerian revolution in behavioral psychology.

The chief theoretical construct of early behaviorists, such as John B. Watson, was the concept of respondent behavior, more commonly designated as reflex behavior.[14] Respondent behavior involves a two-term relationship between a feature of the environment, called the stimulus, and an activity of the organism, dubbed the response. The role of the stimulus was originally conceived as that of 'eliciting' or 'releasing' the response, but in laboratory situations, at least, where such metaphysics may safely be forsworn, one need only assume a functional relationship between the two. The stimulus is merely an occasioning event, one in the presence of which a response is more or less probable. Some S–R associations are natural, such as a dog's salivating at the sight of food, whereas others can be established by conditioning, as Ivan Pavlov demonstrated by getting a dog to salivate also at the sound of a bell.

Examples of respondent behavior abound, both in nature and in the laboratory, and this fact encouraged early behaviorists to promote this simple S–R pattern as the model for all human behavior. With the advantage of hindsight, we now see clearly that there are serious and fundamental limitations of the respondent model, but our less-advantaged predecessors saw only puzzles, to which they responded with a variety of ad hoc devices. Typical of the difficulties with the S–R model is the fact that stimuli that one would expect to be effective often fail to elicit any response whatsoever. More generally, one finds that a variety of responses can be conditioned to the same stimulus, and, that, conversely, a single response can be brought under the control of several different stimuli. None of these phenomena can be explained by a model that speaks only of stimulus and response; something more is needed. It was this situation that led behaviorists such as Edward C. Tolman and Clark Hull to posit 'intervening variables' as hypothetical intermediaries whose function is to account for the failure of a neat one-to-one connection between stimuli and responses.

---

[14] See Watson 1930. The term 'respondent' is Skinner's invention; see Skinner 1938, 20.

From a purely formal point of view, there can be no objection to posited intervening variables – they can be made to do the work expected of them. But from the point of view of the pragmatics of a developing, experimental science of behavior there is much to criticize in a hasty flight to hypothesis. As Skinner noted, the 'central states' that intervening variables are taken to represent invite interpretation as either mental or neurological states, and once such an interpretation is made, central states take on a reality that is more a function of the familiarity of discourse about minds or neurons than of the concrete evidence for these specific posits. Inevitably, one's descriptions of central states borrow heavily from the vocabularies of other disciplines, such as neurophysiology, information theory, or more recently, computing theory and artificial intelligence. Skinner worried that one thus purchases a 'spurious sense of order or rigor' at the expense of experimental control (Skinner 1969, 83).

Recourse to hypothetical intervening variables is especially unwise when the possibility exists of supplementing the basic S–R model with a more directly 'observable' third term, and it is observability, or, more precisely, accessibility to experimental control, that is the chief distinguishing methodological virtue of Skinner's alternative to intervening variables: *contingencies of reinforcement.*

The concept of 'contingencies' encapsulates an important feature of most behavior that had been neglected by many behavioral psychologists up to Skinner's time, namely, the effect that the consequences of present behavior have upon future behavior. In classical, respondent conditioning, reinforcement is provided regardless of the organism's response. Pavlov, for example, did not make the presentation of food contingent upon the dog's salivating; he simply rang the bell and presented the food. In operant conditioning, on the other hand, reinforcement is forthcoming only when a specific kind of response occurs; the rat is fed only when it presses the bar within a suitable length of time after the stimulus is presented.

There is nothing hypothetical about contingencies, especially in well-controlled laboratory situations. The consequences of an experimental organism's behavior are easily inferred from the design of the experimental apparatus. Moreover, the concept of contingencies can, in principle, be called upon to deal with most or all of the problems that the concept of intervening variables was invented to solve.[15]

---

[15] Appreciate the import of the fact that Skinner's reservations about intervening variables concerned mainly (a) the way they are poorly modeled by adventitiously available analogues in other sciences, (b) their, thereby, taking on a spurious reality, and (c) their inaccessibility to experimental control. What one should realize, then, is that his famous opposition to mentalism in psychology was not

In outline, the model that Skinner proposes is that of the *operant*: a three-term functional relationship among stimulus, response, and contingencies of reinforcement. In operant behavior, the probability of an organism's emitting a given response when exposed to a specific stimulus (as measured by the organism's rate of responding in the presence of that stimulus) is primarily a function of the contingencies of reinforcement to which the organism has been subjected.[16] The fact that an operant is a three-term functional relationship has important consequences, for it implies that the identity conditions for operants are not simply those of the component responses. Instead, operants are individuated on the basis of patterns of functional dependence; an operant is not just a response, but a way of responding. Various facile criticisms of Skinner fail precisely because of their neglect of this elementary point, and Skinner's insistence on the functional nature of the relationship constitutive of the operant betrays his debt to the functionalism that Dewey and his students defended a generation earlier.

Over seventy years have passed since the introduction of the operant concept (Skinner 1938), and during that time an abundance of data has been accumulated on the specific ways in which the probability of response depends upon different schedules of reinforcement. Many of the regularities discovered have been found to be independent of the nature of the stimulus, the response, the reinforcer, and even the experimental organism. That fact, together with the success of Skinner and his students in explaining, by means of the operant model, a wide variety of sometimes very complex behaviors, is good reason to continue to promote operant analysis at least as a program for research in psychology. This is not to assert, dogmatically, that operant psychology is the last word in the investigation of human behavior, and, clearly, it is not widely regarded as such today. The point is, instead, to suggest that research according to the operant paradigm has paid, and promises to continue paying, dividends, and that the shortcomings of the operant model are more likely to be discovered through a concerted effort to extend an operant analysis to ever more

driven mainly by general philosophical, operationalist, or positivist scruples about mind or the positing of unobservables. Instead, his worry was the perfectly respectable, scientific worry about the degree of experimental control one has over the details of one's theorizing. Obviously there are intervening variables. Obviously the accumulated experience of the organism inscribes itself somehow in some kind of central states. The problem is one's too quickly adverting to a handy model or some folk-psychological constructs in lieu of first doing the hard work of describing all of the experimentally accessible variables.

[16] I say 'primarily', because other independent variables, such as the organism's state of deprivation, might be relevant under certain circumstances.

complex kinds of behavior than through a critique based upon intuitions about the existence of intentional or mental entities.

Research has shown that operants fall into definite classes defined by the nature of the relationship among stimulus, response, and contingencies of reinforcement. The most important class for the purposes of a behavioral semantics is that of verbal behavior, which Skinner defines as 'behavior reinforced through the mediation of other persons' (Skinner 1957, 2). An example illustrating the crucial property of mediation in a clear form would be a child's acquisition of the verbal operant *Water!*[17] In an appropriately contrived environment, a child might learn to crawl to a dish of water when he or she is thirsty; this would be non-verbal operant behavior, because the response, the crawling, leads directly to the reinforcement. More commonly the child learns to emit a certain vocal response, which, if heard by another member of the verbal community, frequently leads to reinforcement. This is genuine verbal behavior because the reinforcement is contingent upon the response's being, itself, a stimulus for an item of (often non-verbal) behavior by another individual; the latter's response is what reinforces the former's vocal response.

The broad class of verbal operants is subdivided into a number of more specific kinds. *Mands* are exemplified by the child's asking for water. The mand is distinguished, to quote Skinner, 'by the unique relationship between the form of the response and the reinforcement characteristically received in a given verbal community' (1957, 36). To put it crudely, the response 'specifies' its reinforcement. Mands are further distinguished by the fact that no specific stimulus condition is required, except, perhaps, a certain state of deprivation and the presence of an appropriate audience. However, several verbal operants do require a specific kind of stimulus. Some require a specifically verbal stimulus, as suggested by the names Skinner gives them, such as 'echoic behavior', 'textual behavior', 'transcription', and 'intraverbal behavior' (Skinner 1957, chapter 4).

The members of one important class of operants, the *tacts*, do require, however, a non-verbal stimulus. This is the kind of verbal behavior most relevant to a behavioral semantics, because the relationship between response and stimulus in a tact resembles, in some respects, the

---

[17] I follow Skinner's own notational convention – italics and the exclamation mark – for designating the variety of operant, the 'mand', further explained below, of which *Water!* is an instance.

relationship between a word and its referent. In fact, there is an important difference between tacting and referring, one that hints at the larger difference between the study of language and the study of verbal behavior: reference is a two-term relationship between word and object; but the tact, like other operants, is a three-term relationship among response, stimulus, and contingencies of reinforcement. Beyond its requiring a non-verbal stimulus, what distinguishes the tact from other verbal operants is the strength of the control exerted by the prior stimulus – no specific deprivation, audience, or aversive stimulation is required. This is a situation produced by the community's reinforcing the response in question in a wide variety of circumstances, using an assortment of reinforcers, the only common factor being the presence of the appropriate non-verbal stimulus (Skinner 1957, chapter 5). Examples abound: *dog* consistently reinforced in the presence of a dog; *red* consistently reinforced in the presence of red things; and so on.

Given the tact's proximity to the traditional semantic notion of reference, we will want to investigate more carefully what Skinner has to say about this class of operants. But pause, first, to think about some of the important, distinguishing features of Skinner's conception of verbal behavior. First, verbal behavior is not different in kind from other operant behavior, except for its unique, socially mediated mode of reinforcement. The fact that verbal and non-verbal behaviors are thus similar in kind suggests that a Skinnerian behavioral semantics might not have to face a problem that is a continuing source of embarrassment to more traditional philosophies of language, namely, the problem of the relation between language and action. For Skinner, there is no difference in kind between verbal behavior and other behavior. The only possible differences are those that can be characterized in terms of the pattern of functional relationships among stimulus, response, and contingencies of reinforcement.

Second, verbal behavior is defined in an extremely general way, so as to include a wide variety of non-vocal and yet verbal behaviors, some of which we might not intuitively assimilate to language. Any kind of response that is established and maintained by socially mediated reinforcement qualifies as verbal. Among the non-vocal verbal behaviors will be included, of course, writing, typing, signing, pointing, and other forms of signaling, such as Morse code and semaphore, also gesturing, like waving hello and goodbye, indicating assent with a wink and a nod, and so forth. Perhaps more surprising will be the inclusion of behaviors that one might have thought not to be in any way verbal, such as musical, artistic, and athletic performance or a craftsperson's physical manipulation of wood, clay, or paint. To the extent that the reinforcement of Michael Jordan's slam dunk or John Coltrane's improvisation is social, to that extent the behavior is verbal. We often describe athletics, art, and music as forms of expression, even though there might be no obvious semantic content expressed, so we say, with less than full conviction, that what they express must be something like emotional attitudes. But there are no such puzzles from the operant point of view. Such behaviors are expressive in the same way as are other verbal behaviors. Even further removed from intuitions about language might be such examples as the socially mediated reinforcement of the apprentice laboratory assistant's handling of experimental equipment. What it is that might be communicated in training an apprentice is as puzzling as is the expressed content of music or athletic performance. Perhaps, in desperation, we speak of tacit knowledge. But here, too, recognition of the crucial role of social mediation in reinforcement makes clear the continuity with other verbal behavior. Learning to speak and learning to titrate do have a lot in common.

Especially interesting among the class of non-vocal verbal behaviors is what Skinner terms 'subvocal' verbal behavior, which comprises everything from talking to oneself to what some might term 'the language of thought'. How language shapes 'thinking' is a famous problem in psychology.[18] Make language and thought different in kind and you make the problem a hard one. Make thought just a species of subvocal verbal behavior and you make the problem much easier to solve. The point is not that all cognition is just interiorized verbal behavior. That is, obviously, false. The point is just that *much* of what we term 'thinking' is like in kind – in operant terms – to verbal behavior in other forms, even if the responses are non-vocal. What is crucial is, again, the socially mediated way in which those habits of responding are established and maintained.

Finally, from an operant point of view, verbal behavior is, as already emphasized, necessarily social behavior. The distinctively verbal pattern of relations among stimulus, response, and reinforcement is displayed only by an organism that is a member of a verbal community, that is, a group of organisms some of whose behavior is reinforced, not by the natural environment, but by the responses of other members of the group. It follows that there can be no private verbal behavior. This does not rule out verbal behavior under the control of internal stimuli, which previously we would have termed 'talk about private events'; it merely implies that

---

[18] A classic and influential discussion is found in Vygotsky 1934.

such verbal behavior, like all verbal behavior, is shaped and maintained by socially mediated reinforcement.

TACTING AS A SURROGATE FOR REFERRING

As mentioned, the class of verbal operants that Skinner labels 'tacts' affords us the closest behavioral analogue of the semantic notion of reference, this because of the tact's distinguishing characteristic, namely, the control exerted by a prior, non-verbal stimulus. Skinner, himself, says of the tact: 'The resulting control is through the stimulus. A given response "specifies" a given stimulus property. This is the "reference" of semantic theory' (1957, 85). Skinner is not, however, proposing here a behavioral explication of the traditional semantic notion of reference. The referential relationship between a dog and the term 'dog' might be thought to resemble the functional relationship between a dog and the vocal response, *dog*. But, from a behavioral point of view, there are crucial differences, for a term or a word is not at all the same thing as a vocal response, and essential to the tact as a verbal operant is that third term, the history of contingencies of reinforcement – something with no analogue in the reference relationship – for the details of that history are precisely what produce the distinctively strong stimulus control in the tact.

Consider each of these two points of difference in turn. A word or term is not a vocal response. A word or term is an abstract object. A vocal response is a concrete, physiological and physical event. One might think to define a word, behaviorally, as an equivalence or similarity class of utterances and inscriptions not only across a big chunk of the biography of a single individual but also across the careers of all members of a relevant verbal community. A word defined thus as an abstract object might be capable of standing in a reference-like relationship to an object as its referent, and it might even play by the rules of formal syntax and inference. But there is no a priori reason to expect the semantic reference relationship – essentially just a mapping – to supervene upon or reduce to the functional relationship between stimulus and response in the corresponding tact. Nor is there any a priori reason to expect, conversely, that that functional relationship in the tact should track the reference relationship.

The second difference is more important. The tact is a three-term functional relationship among stimulus, response, and contingencies of reinforcement, and such a third term is lacking in the semantic relationship of reference. For Skinner, it is the operant, in this case the tact, that is the proper unit of analysis, not the response alone, even less so the word.

The mistake of taking the response or the word as the proper unit of analysis is labeled the 'formalistic fallacy' by Skinner, who remarks that it is
'common in linguistics and psycholinguistics' and that it is 'most dam-
aging when verbal behavior is analyzed as if it were generated through
the application of rules' (1969, 89–90). It is a mistake precisely because
it directs attention away from the crucial controlling variables in verbal
behavior, foremost among them the contingencies of reinforcement.

There is special irony in neglecting the controlling variables in the
case of tacting as a behavioral surrogate for reference, since, as noted, it
is precisely the contingencies of reinforcement that establish the characteristically strong stimulus control in the tact, the feature of the tact that
makes it seem so much like reference. Such strong stimulus control is
established by contingencies that reinforce the relevant response – *dog* – in
the presence of a discriminative stimulus – a dog – over a wide range of
background conditions and in the absence of any specific deprivation or
aversive stimulation. Change those contingencies and the topographically
identical response can be part of an entirely different verbal operant, such
as a mand.

Skinner devotes a long section of *Verbal Behavior* to what he terms 'The
Problem of Reference' (Skinner 1957, 114–129). He stresses that the tact of
operant theory is not an *explication* of the semantic notion of reference
but a *replacement* for it: 'We are interested in finding terms, not to take
traditional places, but to deal with a traditional subject matter' (115). How
are we to 'deal with' this traditional subject matter? Skinner explains:

Semantic theory is often confined to the relation between response and stimulus which prevails in the verbal operant called the tact. Words, parts of words,
or groups of words on the one hand and things, parts of things, or groups of
things on the other stand in a relation to one another called 'reference', 'denotation', or 'designation'. The relation may be as empty as a logical convention or
it may provide for the 'intention' of the speaker. But how a word 'stands for' a
thing or 'means' what the speaker intends to say or 'communicates' some condition of a thing to a listener has never been satisfactorily established. The notion
of the verbal operant brings such relations within the scope of natural science.
(114–115)

Skinner surveys several noteworthy ways in which the operant surrogate
for reference redirects our understanding of the relation between language
and meaning or verbal behavior and world. One concerns the reference of
general and abstract terms. One might think this a problem in a behavioral
analysis, since reinforcement always takes place in the presence of specific,
concrete objects. But what is required is simply that one systematically

vary stimulus properties noting those in which a specific response is present or absent. What do we find?

We cannot solve this problem by giving the relevant property a sort of object-status as a 'concept' or 'abstraction' – by saying that the response *red* refers to the 'concept of red' or to the 'redness' of something. We never reinforce a response when a 'concept' is present; what is present is a particular stimulus. The referent of an abstract tact, if this term has any meaning at all, is the property or set of properties upon which reinforcement has been contingent and which therefore control the response. We might say that the referent is the *class* of stimuli defined by such a property or properties, but there is little reason to prefer classes to properties. The property correlated with reinforcement must be specified, in physical terms, if we are to remain within the framework of empirical science. (117)

The class of stimuli or the properties thus identified behaviorally as the referent of an abstract term will depend upon the details of the reinforcing practices of the relevant verbal community and might not correspond closely to the philosopher's preanalytic intuitions about natural kinds. Stay with color terms. One can imagine verbal communities wherein the contingencies of reinforcement depend more strongly upon intensity than wavelength, just as one can imagine a community in which, thanks to accidents of plumage among birds that are targets of prey, it is functional to reinforce a single response to colors as different, to us, as red is from purple.

Concrete but general terms might seem to pose fewer puzzles. But here, too, the lesson stands that responses are reinforced always in the presence of particular stimuli. 'Where the stimulus appears to be an object, the object is taken as the referent of the response; yet there is always an element of abstraction. We cannot point to a single chair which is the referent of the response *chair*' (117). From a behavioral point of view, there is a continuum of cases running from proper names to abstract terms, the various tacts differing only in the degree of stimulus generalization involved.

It is a matter of special importance to Skinner that a single response can be under the control of different stimuli, and that a single stimulus can control a multiplicity of responses. Both cases afford a continuum of possibilities. A single response under the control of similar stimuli corresponds to either abstraction or metaphor depending upon how we regard the dimensions along which the stimuli vary. Both begin as examples of what Skinner terms 'extension', which is characterized by a response's being emitted in the presence of stimuli before which the response had not previously been reinforced. Abstraction is a matter of 'generic extension', the variety of extension in which the property making the stimulus effective is that upon which the community's reinforcement was contingent, as when a subject vocalizes *chair* when confronted by a new kind of chair. 'Metaphorical extension', in Skinner's parlance, occurs when the effective property of the stimulus is one that happened to have been present when the response was reinforced but was not among the properties respected by the verbal community's contingencies. Skinner gives the example of a child who, when drinking soda water for the first time, said, *it tastes like my foot's asleep* (92). At the far end of the spectrum, where the stimuli controlling a single response are genuinely different, we have a case of homonymy.

The complementary case of different responses being under the control of a single stimulus represents a case of synonymy, or the behavioral surrogate for synonymy, since the identity resides not in an abstract realm of meaning but in the sameness of the physical conditions defining the stimulus. The synonymy is partial if the stimuli in whose presence the different responses are emitted are similar but not identical. Here, too, a continuum of intermediate cases is found.

Mention of synonymy of course brings to mind the much-contested question of the analytic–synthetic distinction, for one famous line of defense of the distinction seeks to define analytic sentences as ones in the case of which substitution of synonyms for synonyms turns such a statement into a trivial logical truth or a trivial statement of identity. Quine's equally famous objection to this defense is to point out that the definition of 'synonymous' requires, in turn, the notion of 'is analytic' (Quine 1951). One would expect a behaviorist to be skeptical of a principled analytic–synthetic distinction. Do we have here, instead, a free-standing, non-question-begging behavioral definition of synonymy, hence a non-circular definition of analyticity, and, thus, an unexpected Skinnerian defense of the analytic–synthetic distinction?[19]

[19] The notion of 'stimulus synonymy' in Quine's work is related, but reflects Quine's making dispositions to assent to, and dissent from, queries the chief form of behavioral evidence in translation. The sentence is Quine's basic unit of analysis. Stimulus synonymy of sentences, for Quine, is identity of 'stimulus meaning' (Quine 1960, 46). He defines the stimulus meaning of a sentence as the ordered pair of the stimulus conditions under which a subject would assent to a sentence when queried and the stimulus conditions under which that subject would dissent to the sentence when queried (Quine 1960, 32–33). Whether Quine's notion of stimulus meaning and stimulus synonymy admits a straightforward translation into Skinner's theoretical vocabulary is not obvious, however much the two approaches share a behavioral orientation and a commitment to empiricism about language. But 'bachelor' and 'unmarried male' count as stimulus synonymous for Quine, just as *bachelor* and *unmarried male* count as synonymous for Skinner as responses conditioned to the same stimulus.

Skinner, himself, touches upon the question of the analytic–synthetic distinction very near the end of the aforementioned section on 'The Problem of Reference', and he there suggests a very different behavioral surrogate for analyticity. Skinner remarks that one might be tempted to map the analytic–synthetic distinction onto the distinction between the two classes of verbal operants, the intraverbals and the tacts. Intraverbals were mentioned above. Their distinguishing feature, as the name is meant to convey, is that the stimuli are, themselves, instances of verbal behavior, and there is no point-by-point correspondence between stimulus and response, as there is in, say, echoic behavior. Intraverbals come in many forms, including 'chaining', 'word association', and 'translation'.

Most relevant to the question of the analytic–synthetic distinction are intraverbal operants such as emitting the response *four* to the verbal stimulus *two plus two*, or the response *unmarried male* to the stimulus *bachelor*. One might have thought that one says *two plus two is four* because the terms, 'two plus two' and 'four' both refer to the same number, or that one says *a bachelor is an unmarried male* because the terms 'bachelor' and 'unmarried male' refer to all and only the same individuals. All four of these terms are associated with verbal responses that are part of tacts. But Skinner's point is that in a more complicated verbal response like *two plus two is four*, the stronger stimulus control over the utterance *four* is most likely exerted not by the property of the stimulus conditions under which *four* was first reinforced – stimulus conditions almost surely absent in the typical circumstances under which the longer utterance is emitted, such as a classroom mathematics exercise – but, instead, by the prior utterance, *two plus two*. The point is that the response *four* has been separately conditioned both to sets of four things, like four fingers, and to a wide array of verbal stimuli, including the responses *the Gospels* and *sides of a square*. As Skinner explains later in *Verbal Behavior*, the sense of 'certainty' attaching to utterances such as *two plus two is four* is nothing more than the felt strength of the verbal stimulus control, or, in other words, the felt high probability of the response, *four*, on the occasion of the stimulus, *two plus two*.

As should be obvious, however, the cases of *two plus two is four* and *a bachelor is an unmarried male* differ little, if at all, from a behavioral point of view, from cases of seemingly synthetic sentences such as 'Caesar crossed the Rubicon.' In the long history of the community's verbal practices, there once was a time when the response *Caesar* was conditioned to the individual so named. But that time being long in the past, the response *Caesar* is today part of a tact, if at all, only with an image as the stimulus, and even then the far stronger stimulus control is in the form of verbal prompts, such as *the Roman emperor assassinated by Brutus*. In the verbal behavior of most speakers of English, the response, *crossed the Rubicon* is almost exclusively under the stimulus control of the verbal response, *Caesar*. From a behavioral point of view, such 'analytic' and 'synthetic' statements are much alike in point of functional relations among stimulus, response, and contingencies of reinforcement. As Skinner says, 'although all analytic sentences may be intraverbal – and hence have no "referents" in terms of the present relation – all synthetic statements are not necessarily tacts' (Skinner 1957, 129). For Skinner, there is a difference between control by verbal and non-verbal stimuli, hence between intraverbals and tacts. But verbal behavior being the complicated thing that it is, virtually every response is, simultaneously, part of more than one verbal operant, and the only relevant distinction is one of degree, specifically the strength its being conditioned to a verbal or non-verbal stimulus.[20]

A final point of considerable importance to Skinner with respect to the tact as the behavioral surrogate for reference is implicit in what has just been said about the analytic–synthetic distinction. It is that a behavioral analysis does not respect an a priori distinction between sentences and words or even one between words and their parts. For Skinner, the unit of analysis is, again, the verbal operant, which is a functional relationship among stimulus, response, and contingencies of reinforcement. From this point of view, a response can be as long as a whole sentence or as short as a fragment of a word. A sentence-length response, *day is done*, can be conditioned to the setting of the sun as can be the word-length response, *sunset*. Word fragments can be part of tacts. Skinner mentions the example of the initial *sp* in *spit, speak,* and *spew* (one might add *spark, spend,* and *spurt*), all of which are under the partial control of stimuli that include the emission of something, and the terminal *each* in *screech, preach,* and *teach*, all of which are under the partial control of noises produced vocally. Obviously not every response beginning with *sp* or ending with *each* shares these kinds of stimulus control – think of *sparrow* and *reach*. Nor would the partial operant autonomy of *sp* and *each* mean that responses

---

[20] Compare Skinner's interpretation of analytic statements as intraverbals with Quine's notion of 'stimulus analyticity'. For Quine, a sentence is stimulus analytic if he would assent to it after every stimulation (Quine 1960, 55). Here, too, there is no direct translation from Quine's behavioral linguistics into Skinner's theory of verbal behavior, but both Skinner and Quine are pointing to a class of sentences or utterances that exhibit a specific kind of indifference to non-verbal stimuli, and for both, such a difference as there might be between such cases and those exhibiting greater dependence upon non-verbal stimuli is one of degree, not kind.

like *speech* are properly analyzed simply as composites, since clearly the *it* in *spit* is not part of a similarly autonomous functional unit, and the whole response *speech* is obviously under determinate stimulus control as part of a tact.

A complete analysis of an episode of verbal behavior would identify multiple verbal operants at the level of the sentence, the phrase, the word, and the syllable, with no one such functional element uniquely suited to serve as a carrier of meaning. But a more traditional semantics demands a principled distinction: The word or the phrase refers, but it is the sentence that is the bearer of truth-values. Words and phrases refer, but it is sentences that assert. Can such distinctions be reflected in a behavioral analysis.

Skinner's view of assertion involves another class of verbal operants, the *autoclitics*. The distinguishing feature of autoclitics is that the controlling stimuli are properties of other verbal operants. One such property might be the strength of a response, which is to say, the probability of its being emitted in given stimulus conditions. Think of the way one expresses doubt by rising inflection on a word or phrase whose applicability is in question. I see someone resembling my nephew, John, but the context might be one in which John is not expected, or my view might be partially obscured, and so I utter the name *John* with rising pitch. What I might think is uncertainty is, from a behavioral point of view, just a comparatively weak probability of response, and it is that weakness of the probability of responding that functions as the stimulus controlling the rising inflection.

In English the weight of assertion is borne mainly by the forms of the verb 'to be'. Consider the first- and third-person present form, 'is'. Consider now two cases. In the first, I am not sure whether the person I see is my nephew. I utter *John* with rising inflection. Then I get a clearer view and I utter *it IS John*. For Skinner, *is* is mainly what he terms an assertive autoclitic.

Another property of verbal behavior that might figure in an accompanying assertive autoclitic is the nature of the verbal operant in question. Skinner writes:

If I know that someone has said wolf and nothing else, the response will be of very little use. The speaker may be calling for help, describing an animal at the zoo, reading a sign, repeating what he has heard, or completing the phrase Big, bad ... An autoclitic will sharpen the effect by indicating some of the sources of strength, as well as the degree of strength [in both stimulus control and strength of response]. The assertive autoclitic has the specific function of indicating that the response is emitted as a tact or, under certain circumstances, as an interverbal. (Skinner 1957, 327; ellipsis in the original)

From the operant point of view, assertion has nothing to do with the linguist's distinction between word and sentence. The assertive autoclitic is but another, distinctive pattern of functional relations among stimulus, response, and contingencies of reinforcement. The word–sentence distinction is irrelevant. Again, Skinner speaks directly to the philosopher's worry:

It is sometimes said that the word is inanimate but that language comes to life in the sentence. Words by themselves say nothing; it is the sentence which asserts. This is not the present distinction. (Skinner 1957, 327)

What, then, of truth? In traditional philosophical semantics, the reference of words undergirds the truth of the sentences composed with those words. 'Caesar crossed the Rubicon' is true because 'Caesar' refers to Caesar, 'the Rubicon' refers to the Rubicon, and Caesar passed over the Rubicon in the manner indicated by the verb 'crossed'. For Skinner, the semantic notion of truth is replaced by the strength of the correlation between stimulus and response in the tact. In an operant analysis of verbal behavior, the only resources are the functional relations among stimulus, response, and contingencies of reinforcement. Of those, the strength of the correlation between stimulus and response in the tact is the best surrogate for the semantic notion of truth.

Verbal behavior is, by definition, social in the sense that the reinforcement of verbal responses is socially mediated. The speaker is, perforce, a member of a verbal community. Over a broad range of circumstances, it makes obvious good sense for the community to maintain contingencies of reinforcement that maximize the correlation between stimulus and response in the tact, which is to say, the differential strength of the stimulus control and the probability of response. Everything from mundane, day-to-day community life up to the long-term survival of the community is facilitated by predictable and reliable regularity in a speaker's responding verbally, the same way, to the same stimulus over a wide variety of circumstances.

The key to maintaining such strong correlations is generalized reinforcement. The community reinforces the response *lion* always and only when the speaker is in the presence of a lion or such evidence of a lion's presence as a roar, a scent, or a certain kind of movement in the grass. Background conditions are allowed to vary as are the circumstances of the speaker. In Skinner's own words:

Generalized reinforcement is the key to successful practical and scientific discourse. It brings the speaker's behavior most narrowly under the control of the current environment and permits the listener to react to that behavior most successfully in lieu of direct contact with the environment. When the correspondence with a stimulating situation is sharply maintained, when the listener's inferences regarding the objective situation are most reliable, we call the response 'objective', 'valid', 'true', or 'correct'. (Skinner 1957, 147)

But, as Skinner also notes, stimulus control is never perfect, nor is verbal behavior ever completely independent of the speaker's current circumstance. One can contrive contingencies that maximize stimulus control and minimize the effects of bias, inattention, and fatigue. Still, the functional correlation between stimulus and response that is the behavioral surrogate for truth is never perfect and always a matter of degree.

Truth as strength of correlation between stimulus and response in the tact is the best behavioral surrogate for the semantic notion of truth. Truth as correlation between stimulus and response might seem like truth as correspondence, but the two differ for the same reasons that tacting is a replacement for, not an explication of, referring: A word is not a response and a tact is a three-term functional relationship among stimulus, response, and contingencies of reinforcement, whereas truth is a two-term relationship between statement and fact. And, as with reference, so, too, with truth: The contingencies of reinforcement make all the difference, since it is the generalized nature of the reinforcement that ensures the strong nature of the correlation between stimulus and response.

FURTHER CONSEQUENCES OF THE TACT AS A BEHAVIORAL
SURROGATE FOR REFERENCE

Causal theories of reference were discussed, above, as a form of naturalism about reference. In them, an originary naming or dubbing or anointing – 'This is Caesar' – fixes reference once and for all and always. How does an operant naturalism compare? Skinner discusses this very question, though not as a point about a causal theory of reference:

When we say that the word *Caesar* refers to Caesar, dead though he has been these thousand years, we are clearly not talking about the behavior of a contemporary speaker. A response of this form is almost certainly intraverbal, if it is not textual or echoic. A process of educational reinforcement has brought it under the control of various sets of verbal circumstances. Theoretically we should be able to trace these circumstances back to an instant in which a response was made to Caesar the man. The study of history assumes valid chains of this sort, and a predilection for primary sources is essentially the avoidance of unduly long, and hence probably faulty, chains. But the verbal behavior of the modern historian is still mostly intraverbal. (Skinner 1957, 128–129)

For Skinner, the essential point is not a causal relationship between *Caesar* and Caesar in an originary act of naming, but rather the controlling contingencies of reinforcement both back then and now. Back then what was crucial was the strength of the correlation between Caesar as stimulus and *Caesar* as response. Causality was involved, to be sure, but even then the causal chains were long and complicated, and the details of those causal chains were ultimately irrelevant to the brute fact of the strong stimulus control of Caesar on *Caesar*. But today the relevant contingencies are purely intraverbal (or textual or echoic). The reinforcing practices of ancestral verbal communities are relevant to understanding the genealogy of the reinforcing practices of contemporary verbal communities. But what matters most today is the status of *Caesar* as an intraverbal. There is, then, a kernel of truth in the causal theory of reference, judged from the point of view of an operant analysis. But it is only that operant analysis that makes it clear when, where, and how the causal connection between *Caesar* and Caesar was once minimally relevant, and that same operant analysis exhibits the vastly diminished relevance of that causal connection for appreciating the behavioral semantics of *Caesar* today.

What does an operant analysis of verbal behavior teach us about what Quine once termed the 'vagaries of reference' (Quine 1960, 125): vagueness, ambiguity, referential opacity, and referential inscrutability? Vagueness and ambiguity are straightforwardly transcribed into questions about varieties of indefiniteness in the controlling stimuli in the tact. Referential inscrutability is more interesting. It is the analogue for referring terms of the indeterminacy of radical translation with respect to sentences. Just as behavioral evidence in the form of dispositions to assent to and dissent from queries systematically underdetermines our translation of 'Gavagai' as 'Rabbit' (or, as Quine notes, 'Lo, a rabbit' – sentences are the unit of analysis), 'Undetached rabbit part' or 'Temporal slice of a rabbit', so too the behavioral evidence underdetermines the reference of the term 'gavagai', leaving rabbits, undetached rabbit parts, and temporal slices of rabbits as equally serviceable candidates.

The surrogate question for Skinner would have to be one about the controlling variables in tacts comprising *gavagai* as the response among members of the native verbal community and the relationships of those functional patterns to the controlling variables in corresponding tacts among the members of the linguist's verbal community, including those that comprise the responses *rabbit, undetached rabbit part*, and *temporal slice of a rabbit*. The operant surrogate for the thesis of referential inscrutability would then be simply the mundane fact that the controlling stimulus in each of those tacts could equally well be described physically as rabbit, undetached rabbit part, or temporal slice of a rabbit, and, conversely, that each of the three responses – *rabbit, undetached rabbit part,* and *temporal slice of a rabbit* – could equally well be conditioned to those stimuli thus described. What was a surprising and provocative thesis about reference when first propounded by Quine becomes a triviality within an operant analysis of verbal behavior.

But an operant analysis also affords tools for understanding why the thesis of referential inscrutability (and translational indeterminacy) is so surprising. It is because, from an operant point of view, all of those responses – *rabbit, undetached rabbit part,* and *temporal slice of a rabbit* – also figure in a wide array of other verbal operants, including mands, intraverbals, textuals, echoics, and others, and it is in the functional relationships defining those other operants where the obvious and salient differences among *rabbit, undetached rabbit part,* and *temporal slice of a rabbit* are to be found. They are different responses, but the differences are largely irrelevant to their status within the crucial tacts, hence irrelevant from the point of view of the behavioral surrogate for reference.

Step back, now, from the immediate context of theories of reference to other contexts in which one wants to deploy a theory of reference. There are many such, but focus now only on epistemology and philosophy of science. If we think that knowledge is expressed in language, then an operant analysis of verbal behavior should make a difference in the way we theorize knowledge. As it happens, an operant analysis promises significant advantages on many fronts. Consider for now just these two, both of which concern areas of significant activity in the recent literature.

Philosophers of experiment argue that the philosopher of science's traditional emphasis on theory neglected the cognitive content that lives in experimental practice (see, for example, Franklin 1986). But the puzzle has always been how to understand especially the kind of non-discursive knowledge that is so much a part of experimental practice. To call it tacit knowledge is just to invent a name for our puzzlement. The operant approach gives us an answer. Since the experimental practices in question most definitely comprise behaviors whose governing contingencies are socially mediated, then such practices represent verbal behavior in exactly the same way in which discursive or vocal or textual behaviors do. In the domain of experimental practice, one will even find examples of intra-verbals – manual verbal responses under the stimulus control of other manual verbal responses – echoics, and mands. Some crucial aspects of experimental practice will, of course, be non-verbal, the reinforcement coming from nature itself. But in many important ways experimental practice will prove to be just as verbal as publishing in the journals and speaking at professional meetings. From an operant point of view, the continuities between theory and experimental practice are many and obvious.

Social epistemology is another area of major interest in the contemporary epistemology and philosophy-of-science literature (see, for example, Goldman 1999, Longino 1990, and Solomon 2001). Here the puzzle has been to understand how knowledge can live in a community rather than in the mind of an individual. Dewey long ago pointed a way to an answer, with his attack on Cartesian solipsism in *The Quest for Certainty* (Dewey 1929). But while Dewey's psychology was as resolutely social as Skinner's, it lacked the latter's technical sophistication about verbal behavior.

As we have seen, Skinner defines verbal behavior precisely as behavior in which the contingencies of reinforcement are socially mediated. It follows that an epistemology constructed on this foundation will be social by definition. The burden of puzzlement will be reversed. The puzzle will be to understand how inherently social epistemic behaviors can (mistakenly?) be thought to reside in the minds of individual knowers. The solution to that puzzle has, however, also been suggested by Skinner with the above-mentioned notion of 'subvocal verbal behavior'. The idea is that vocal responses can be so attenuated as to be inaudible, even to the speaker, and yet still be controlled by the kind of socially mediated reinforcement definitive of verbal behavior. Thinking about an object is, thus, just a curiously private way of speaking about objects. Even the *Cogito* comes to be seen as a socially maintained response to a complicated set of verbal and non-verbal stimuli.

CONCLUSION: DIDN'T CHOMSKY PROVE SKINNER
JUST PLAIN WRONG?

It was said above that one of the reasons for contemporary neglect of Skinner's operant approach to verbal behavior is the legend of Chomsky's having killed the project with his famous review of *Verbal Behavior* (Chomsky 1959). Chomsky objects to many of Skinner's ideas, faulting the clarity even of the basic notions of stimulus, response, and reinforcement. His objections on these points are unsubtle, unsympathetic, and, in some instances, simply uninformed. Remarkably, he accuses Skinner of simply adopting wholesale the traditional semantics of reference. But one charge proved especially damning. Chomsky argued that Skinner's operant analysis of verbal behavior was, in principle, incapable of explaining the adult speaker's well-documented competence in recognizing the grammaticality of sentences wholly new to the speaker:

These are very remarkable abilities. We constantly read and hear new sequences of words, recognize them as sentences, and understand them. It is easy to show that the new events that we accept and understand as sentences are not related to those with which we are familiar by any simple notion of formal (or semantic or statistical) similarity or identity of grammatical frame. Talk of generalization in this case is entirely pointless and empty. It appears that we recognize a new item as a sentence not because it matches some familiar item in any simple way, but because it is generated by the grammar that each individual has somehow and in some form internalized. And we understand a new sentence, in part, because we are somehow capable of determining the process by which this sentence is derived in this grammar. (Chomsky 1959, 56)

Readers familiar with Chomsky's own then-evolving project of transformational grammar (see, for example, Chomsky 1957) would get the point, a famous one from Chomsky, that only the posit of a universal and internalized grammatical competence can explain the speaker's capacity for recognizing the well-formedness of novel sentences. But readers familiar with Skinner's *Verbal Behavior* will see at once the question-begging nature of the charge. For precisely this point – the speaker's ability to respond to novel sentences as conforming to rules of grammar – is discussed by Skinner at length in chapter 13, 'Grammar and Syntax as Autoclitic Processes' (Skinner 1957, 331–343).

Chomsky discusses this chapter, but dismisses Skinner's approach to grammar with the following glib remark:

It is evident that more is involved in sentence structure than insertion of lexical items in grammatical frames; no approach to language that fails to take these deeper processes into account can possibly achieve much success in accounting for actual linguistic behavior. (Chomsky 1959, 54)

The actual evidence for what is claimed to be 'evident' is simply not provided. A detailed and persuasive rejoinder to Chomsky on this point was provided by Kenneth MacCorquodale (1970, 94–95). The key point missed or, rather, dismissed by Chomsky is simply that elements of syntax become, themselves, controlling stimuli in the production of verbal behavior, and so grammatical verbal behavior is to be explained as are all other elements of verbal behavior. There is, simply, no demonstrated need to posit internalized grammatical rules. Here is how MacCorquodale makes the point:

> In sum, the verbally competent person can discriminate a syntactic dimension in speech as a stimulus, and he can emit speech which has syntactic properties in the sense that a hearer can discriminate them. This does not prove in any way that some underlying theory governs both behaviors. A child learns both to walk and to discriminate walking. Nothing is gained by saying that therefore he has constructed a theory of walking which he uses in his perceptions and in his activities. So he may be conceived to learn to speak and to perceive speech, directly and without stopping to construct a theory or apply a rule. (MacCorquodale 1970, 95)

Chomsky's error is then much like that of other students of behavior who leap to the postulation of intervening variables – in this case an internalized structure of grammatical rules – without realizing that the contingencies of reinforcement can do that work instead.

It is ironic that Chomsky faults Skinner for a failure to explain novelty in verbal behavior, because novelty was one of Skinner's main concerns. It helps to recall that the young Skinner wanted more than anything else to be a poet, even to the extent of trying out the poet's life in Greenwich Village (Skinner 1976). Skinner loved language and had a remarkable ear and eye for its complexity, subtlety, and nuance. Creativity is central. Among the most rewarding and impressive sections of *Verbal Behavior* are those devoted to mapping and explaining novelty in the form of metaphorical extension, stimulus generalization, and response induction. Miss that feature of the program and one misses what was most important to Skinner.

What, then, finally, of reference? As explained, what Skinner provides is not an explication of the traditional semantic notion of reference but a behavioral surrogate for it in the concept of the tact, a class of verbal operants distinguished by the strong control of non-verbal stimuli. As part of Skinner's larger operant analysis of verbal behavior, the tact affords the most sophisticated version of semantic naturalism on offer in the literature. It should be seen as the legacy of Dewey and the more fully formed behavioral semantics toward which Quine was reaching.

REFERENCES

Angell, James R. (1904) *Psychology: An Introductory Study of the Structure and Function of Human Consciousness.* New York: Henry Holt.

Banks, Erik C. (2003) *Ernst Mach's World Elements: A Study in Natural Philosophy.* Dordrecht: Kluwer.

Boyd, Richard (1979) Metaphor and theory change: What is 'metaphor' a metaphor for? In Andrew Ortony (ed.) *Metaphor and Thought.* New York: Cambridge University Press, pp. 481–532.

Bradie, Michael (2008) Evolutionary epistemology. In Edward N. Zalta (ed.) *Stanford Encyclopedia of Philosophy.* http://plato.stanford.edu/entries/epistemology-evolutionary/

Campbell, Donald T. (1974) Evolutionary epistemology. In P. A. Schilpp (ed.) *The Philosophy of Karl R. Popper.* LaSalle, IL: Open Court, pp. 412–463.

Chomsky, Noam (1957) *Syntactic Structures.* The Hague: Mouton.

(1959) Review of *Verbal Behavior*, by B. F. Skinner. *Language* 35: 26–58.

Coffa, Alberto (1991) *The Semantic Tradition from Kant to Carnap: To the Vienna Station.* New York: Cambridge University Press.

Dewey, John (1896) The reflex arc concept in psychology. *Psychological Review* 3: 357–370.

(1910) *The Influence of Darwin on Philosophy and Other Essays in Contemporary Thought.* New York: Henry Holt.

(1922) *Human Nature and Conduct: An Introduction to Social Psychology.* New York: Henry Holt & Co.

(1925) *Experience and Nature.* Chicago and London: Open Court.

(1929) *The Quest for Certainty: A Study of the Relation of Knowledge and Action.* New York: Minton, Balch.

Fodor, Jerry A. (1975) *The Language of Thought.* Cambridge, MA: Harvard University Press.

Franklin, Allan (1986) *The Neglect of Experiment.* Cambridge University Press.

Frege, Gottlob (1892) Über Sinn und Bedeutung. *Zeitschrift für Philosophie und philosophische Kritik* 100: 25–50.

Friedman, Michael (2000) *A Parting of the Ways: Carnap, Cassirer, and Heidegger.* Chicago: Open Court.

Goldman, Alvin (1999) *Knowledge in a Social World.* Oxford University Press.

Hatfield, Gary (1990) *The Natural and the Normative: Theories of Spatial Perception from Kant to Helmholtz.* Cambridge, MA: MIT Press.

(2003) Behaviourism and naturalism. In Thomas Baldwin (ed.) *The Cambridge History of Philosophy, 1870–1945.* Cambridge University Press, pp. 640–648.

Heidelberger, Michael (2004) *Nature from Within: Gustav Theodor Fechner and His Psychophysical Worldview*, trans. Cynthia Klohr. University of Pittsburgh Press.

Hothersall, David (2004) *History of Psychology*, 4th edn. New York: McGraw-Hill.

Jacquette, Dale (ed.) (2003) *Philosophy, Psychology, and Psychologism: Critical and Historical Readings on the Psychological Turn in Philosophy.* Dordrecht: Kluwer.

Katz, Jerrold J. (1972) *Semantic Theory.* New York: Harper & Row.

Katz, Jerrold J. and Fodor, Jerry A. (1963) The structure of a semantic theory. *Language* 39: 170–210.

Kitcher, Philip (1992) The naturalists return. *Philosophical Review* 101: 53–114.

Kripke, Saul (1971) Identity and necessity. In Milton K. Munitz (ed.) *Identity and Individuation.* New York University Press, pp. 135–164.

(1972) Naming and necessity. In Donald Davidson and Gilbert Harman (eds.) *Semantics of Natural Language.* Dordrecht: Reidel, pp. 253–355. Reprinted as *Naming and Necessity.* Cambridge, MA: Harvard University Press, 1980.

Kuhn, Thomas S. (1962) *The Structure of Scientific Revolutions.* University of Chicago Press.

Kusch, Martin (1995) *Psychologism: A Case Study in the Sociology of Philosophical Knowledge.* London and New York: Routledge.

Lee, Harold N. (1973) Dewey and the behavioral theory of meaning. *Tulane Studies in Philosophy* 22: 51–62.

Longino, Helen (1990) *Science as Social Knowledge.* Princeton University Press.

MacCorquodale, Kenneth (1970) On Chomsky's review of Skinner's *Verbal Behavior. Journal of the Experimental Analysis of Behavior* 13: 83–99.

Neisser, Ulrich (1967) *Cognitive Psychology.* New York: Appleton–Century–Crofts.

Putnam, Hilary (1970) Is semantics possible? *Metaphilosophy* 1: 187–201. Revised version reprinted in *Mind, Language, and Reality,* vol. 11 of *Philosophical Papers.* Cambridge University Press, 1975, pp. 139–152.

Quine, W. V. O. (1951) Two dogmas of empiricism. *Philosophical Review* 60: 20–43. Reprinted in *From a Logical Point of View.* Cambridge, MA: Harvard University Press, 1953, pp. 20–46.

(1960) *Word & Object.* Cambridge, MA: MIT Press.

(1968) Ontological relativity. *Journal of Philosophy* 65: 185–212. Reprinted in Quine 1969b, pp. 26–68.

(1969a) Epistemology naturalized. In Quine 1969b, pp. 69–90.

(1969b) *Ontological Relativity and Other Essays.* New York and London: Columbia University Press.

(1973) *The Roots of Reference.* LaSalle, IL: Open Court.

Rorty, Richard (1967) *The Linguistic Turn: Recent Essays in Philosophical Method.* University of Chicago Press.

(1976) Realism and reference. *Monist* 59: 321–337.

Ryle, Gilbert (1949) *The Concept of Mind.* London and New York: Hutchinson's University Library.

Sellars, Roy Wood (1922) *Evolutionary Naturalism.* Chicago and London: Open Court.

Shimony, Abner and Nails, Debra (eds.) (1987) *Naturalistic Epistemology: A Symposium of Two Decades.* Dordrecht: Reidel.

Shook, John R. (2000) *Dewey's Empirical Theory of Knowledge and Reality.* Nashville, TN: Vanderbilt University Press.

Skinner, Burrhus Frederic (1938) *The Behavior of Organisms: An Experimental Analysis.* New York: Appleton–Century–Crofts.

(1957) *Verbal Behavior.* New York: Appleton–Century–Crofts.

(1969) *Contingencies of Reinforcement: A Theoretical Analysis.* New York: Appleton–Century–Crofts.

(1976) *Particulars of My Life: Part One of an Autobiography.* New York: Knopf.

Solomon, Miriam (2001) *Social Empiricism.* Cambridge, MA: MIT Press.

Tarski, Alfred (1935) Der Wahrheitsbegriff in den formalisierten Sprachen. *Studia Philosophica* 1: 261–405. Translated by J. H. Woodger as The concept of truth in formalized languages. In *Logic, Semantics, Metamathematics: Papers from 1923 to 1938.* Oxford: Clarendon Press, 1956, pp. 152–278.

Titchener, Edward B. (1896) *An Outline of Psychology.* New York and London: Macmillan.

Vygotsky, Lev S. (1934) *Myshlenie i rech.* Moscow: Gosudarstvennoe sotsial'no-ekonomicheskoe izdatel'stvo [State Social and Economic Press]. Translated by Eugenia Hanfmann and Gertrude Vakar as *Thought and Language.* Cambridge, MA: MIT Press, 1962.

Watson, John B. (1930) *Behaviorism,* rev. edn. University of Chicago Press.

Wittgenstein, Ludwig (1922) *Tractatus Logico-Philosophicus.* London: Kegan Paul, Trench, and Trübner.

CHAPTER 10

Causal descriptivism and the reference of theoretical terms

Stathis Psillos

I INTRODUCTION

An adequate theory of reference of theoretical terms should satisfy two important conditions.

I. The burden of reference of theoretical terms rests with theory in the sense that what they refer to is determined (at least to a large extent) by the theory in which they feature prominently.
II. Theoretical terms are transtheoretical in the sense that they can refer to the same entity even though they may occur in different theories (or, similarly, different terms may refer to the same entity).

There are many reasons why this should be so. When it comes to (I), it should be obvious that unlike many ordinary objects, theoretical entities cannot be pointed at, perceived by the naked eye, presented to our sensory modalities, and the like. If anything, we have causal contact with them by virtue of their effects; and cognitive contact with them, by virtue either of our causal contact with them or of the confirmation of theories that are about them (or both). Even though causal contact with theoretical entities is not theoretical *in itself*, ascertaining it requires or relies upon theoretical knowledge (or at least beliefs), since even if the causal relation is not itself theoretical, what causes certain effects (and hence what it is that we have causal and cognitive access to) should be theoretically identified by its properties. When it comes to (II), unless we take seriously the possibility that each and every theory that emerges presents us with an image of the world, fresh and totally unrelated to whatever has gone on before, there must be referential continuity between succeeding theories as a necessary condition for being able to talk about the same entities, even though different things may be said of them by different theories. Condition (II) is necessary for progress, for blocking incommensurability, and for developing a fairly realist image of the world.

It might be protested, right from the outset, that these conditions are tailored to a realist approach to science. This is only partly true. They are indeed significant for anyone who takes scientific theories literally – for semantic realists, let us say. Clearly, those who take it that theoretical terms are cognitively insignificant need not bother with (I) or (II). But one need not subscribe to strong realist views to adopt (I) and (II). In particular, one need not accept the view that science does succeed in delivering true theories of the world to accept (I) and (II). Do bear in mind that (I) and (II) are semantic and not epistemological theses – though they can certainly help the realist epistemological cause. Be that as it may, my point of view will be a realist one. For I take very seriously Putnam's (1962) 'short argument' for theoretical terms, namely that the reason that theoretical terms are necessary is precisely that scientists employ terms like 'electron', 'virus', 'space-time curvature', and so on – and advance relevant theories – because they wish to *talk about* electrons, viruses, the curvature of space-time, and so on; that is, they want to find out about the deep structure of the world. It is the theoretical terms that provide scientists with the necessary linguistic tools for talking about things they *want* to talk about.

A tension, however, appears between (I) and (II). The standard descriptivist theories of reference (at least as they are assumed in the philosophy of science) satisfy (I) but not (II), while the standard causal theories of reference satisfy (II) but not (I).

The tragedy of the descriptivist theories of reference, when applied to scientific theories, is that while they honour the thought that theoretical terms do have factual reference (this being whatever satisfies certain theoretical descriptions), they face a genuine difficulty in honouring the thought that terms featuring in distinct theories have the same reference. But, I will argue, the culprit is *not* the idea that reference is fixed by descriptions – or better, that descriptions do play a significant role in reference-fixing. The culprit is *holism* – the view that the reference is fixed by the whole network of theoretical statements that a term is part of.

The tragedy of the causal theories of reference, when applied to scientific theories, is that while they honour the thought that the same (or different) terms that feature in different theories may well refer to the same worldly entity (this being the actual cause of whatever phenomena led to the introduction of a new theoretical entity and a term to refer to it), they face a genuine difficulty in explaining referential failure – insofar as there are causes of certain phenomena Φ, *any* term in *any* theory devised to stand for the cause of Φ does stand for these causes, irrespective of how wrong might be the theoretical descriptions associated with this term.

This is but an instance of a general failure of the causal theories to grant a genuine role to theories in reference-fixing. But, the culprit, I shall argue, is not the idea that causation is involved in the mechanism of reference-fixing. The culprit is the thought that the bare causal relation *in itself* is enough to fix the reference of theoretical terms.

After discussing the problems faced by the two standard theories of reference, as they are applied to the issue of the reference of theoretical terms, I will advance and defend causal descriptivism as an alternative account of the reference of the theory-dependent terms. I will argue that a hybrid theory of reference of theoretical terms (a) is independently well-motivated; and (b) meets in a satisfactory way conditions (I) and (II) above. Finally, I will unravel the key difficulties of a recent attempt to (dis)solve the issue of the reference of theoretical terms by recourse to Ramsey sentences.

2   THE SWING OF THE PENDULUM I: DESCRIPTIVIST THEORIES OF REFERENCE

The story of the descriptivist theories of reference is well-known, so the reader can be spared a lot of details and subtleties. For present purposes, it is enough to recapitulate some of their key ideas. The characterization 'descriptivist *theories*' is not accidental, since the key ideas have undertaken considerable modifications over the years and in light of important philosophical controversies. The central thought, however, is that the competent speaker of a language (or the competent user of an expression) must know some identifying facts about the referent in order to refer successfully to it, where these identifying facts are captured by descriptions of the referent. If nothing satisfies the associated descriptions, then the expression does not refer to anything. But for an expression to refer to anything at all, it must be the case that its associated descriptions must be satisfied; that is, there must be something that has the properties attributed to it by the associated descriptions. It is useful to think of the associated descriptions as Fregean *modes of presentation*: the referent is presented as being in a certain way (and the same referent might be presented in different ways). We can even think of the modes of presentation as the *senses* of expressions; hence, senses are descriptive mechanisms that fix reference. That sense determines reference is then an important plank of descriptivist theories: an expression acquires its reference (if any) via its sense.

Until Kripke's (1980) well-known attack on the descriptivist theories, they were the only game in town. Though they were never used *explicitly* as theories of reference-fixing in the philosophy of science, they were quietly operating in the background. For instance, the very (but short-lived) idea of providing explicit definitions of theoretical terms was but an application of the descriptivist theory of reference coupled with the idea that the associated descriptions should be couched in an observational vocabulary assumed to be independently meaningful and semantically kosher. Even more relaxed approaches to the meaning of theoretical terms – e.g. those based on Carnapian (1936) reduction sentences – were assuming the descriptivist theories. The key problem, however, that these looser approaches faced – namely that theoretical terms could not, in the end, be exhaustively defined by means of logical operations on observational statements – brought to light a shortcoming of the coupling of descriptivist theories with the allegedly sharp dichotomy between observational terms and theoretical ones. The problem was in the *coupling*, of course, since there is nothing in the descriptivist theories themselves that forbids the referent of an expression to be an unobservable entity. What matters is whether the associated descriptions are satisfied or not, not whether the satisfier can or cannot be described in a supposed privileged observational vocabulary. Indeed, the demise of the distinction between observational and theoretical terms (that is, the demise of the programme of providing distinct semantics for these allegedly distinct types of term) freed the descriptivist theories of a burden. The central thought that the burden of reference rests upon descriptive phrases was freed from the redundant and ill-motivated extra condition that these descriptive phrases should be couched in an observational vocabulary.

The way forward was to treat all vocabulary of scientific theories on a par and to claim that the descriptive phrases that fix the reference of theoretical terms are supplied by the theory as a whole: the meaning of a term is determined by its relations to other terms. More specifically, a seminal idea – due to Putnam – was that all theoretical concepts of a theory are 'law-cluster' concepts: they get their meaning via the plethora of nomological statements in which they occur. This view has two immediate consequences. The first is that since the nomological statements that constitute the 'law cluster' are synthetic, there is no way to separate them into two cleanly divided camps: those that fix the meaning of a concept and those that specify its empirical content. All do both. Hence, there is no way to draw the analytic–synthetic distinction *within* a theory. The second consequence is that the meaning (and hence the reference) of theoretical terms is fixed in a holistic way. The theoretical terms are implicitly defined, as it were, by the nomological statements that specify the network of their connections within a theory – and, more specifically, by *all* of these nomological statements.

Semantic holism contributed significantly to the wide acceptance of the claim that theoretical discourse is meaningful and that theoretical terms have putative factual reference. But there are two shortcomings. The first has to do with the very idea of implicit definition. The second has to do with the extent of holism.

The chief attraction of implicit definitions is precisely that they fix meaning, not by analysing already known and understood concepts, but by legislating in a stipulative manner the truth of certain propositions, of which the defined concepts are constituents. Hence, they *create* or *constitute* meanings: for something to be an F (that is, for the concept of F to be applicable to it), such and such conditions must be satisfied. The chief drawback, however, is that an implicit definition is a kind of *indefinite description*: it defines a whole class of (or classes of) objects which can realize the formal structure, as defined by a set of axioms. There is no straightforward way in which unique satisfaction of the descriptions associated with the implicitly defined theoretical terms can be assumed. It might be thought that this latter problem is avoided at the point of application of the theory to the empirical world – at least this is what Schlick thought ([1932] 1979). But this is too quick. *First*, for it to work in the first place, the observational vocabulary (or a vocabulary fit to describe the empirical content of the theory) should be antecedently given and independent of the theory. If the meaning of this vocabulary too is fixed by the theory (that is, by the very same implicit definition that fixes the meanings of theoretical terms), then it can no longer anchor the theory to the empirical world. *Second*, even if the observational vocabulary is antecedently given and fixed independently of theory, it is still an open possibility that the theoretical structure that is fixed by the implicit definition is multiply realized.

We shall discuss these problems in more detail in Section 5. For the time being, let me focus on the issue of the extent of meaning holism. Here the otherwise liberating coupling of descriptivist theories with semantic holism becomes an unstable position. Descriptivist theories do warrant attributing reference to theoretical terms, but if semantic holism is rampant, the meaning of *all* terms (including those that normally count as observational) is specified in a holistic way. If semantic holism is moderate (if, that is, some terms get their meaning in a non-holistic way), a story needs to be told as to how this is possible which does not commit to semantic double standards. This story might well be possible.[1] Whether or not semantic holism is rampant, the troublesome consequence is that every time the theory changes, the meanings of *all* terms whose meaning is determined by the theory change too. We have then a thesis of radical meaning variance in theory change. Coupled with descriptivist theories of reference, and their key point that sense determines reference, semantic holism yields as a further consequence an even more radical thesis, namely reference variance.

The view that change of reference is *inevitable* when theories change – and hence the denial of the view that theoretical terms are transtheoretic – has been associated with Kuhn and Feyerabend. One of the standard examples concerns the terms 'mass', 'space', and 'time' as they occur in Newton's, and then in Einstein's, theory. As Kuhn ([1962] 1970, 102) famously stated:

the physical referents of these Einsteinian concepts [i.e. mass, space, and time] are by no means identical with those of the Newtonian concepts that bear the same name.

Equally famously, Kuhn ([1962] 1970, 128) went on to suggest that *observational terms* too, such as 'planet', have different meaning as well as reference when they occur in different theories (e.g. Ptolemy's and Copernicus').

This bizarre conclusion is, of course, the outcome of a dual commitment to semantic holism *and* the descriptivist theory of reference. Semantic holism entails that if there are changes in the theoretical/inferential network in which a term is embedded, the meaning of the term necessarily changes; the *new* meaning of the term is then given by its function within the *new* network.

Note that the descriptivist theories of reference are not *necessarily* holistic. Even when more than one description is associated with a term, it is allowed that individuals (the referents) are picked out by *weighted* descriptions. Then, not each and every change in the cluster of descriptions associated with the terms yields reference variance. Besides, descriptivist theories allow that two different descriptions may pick out the *same* individual (i.e. they may be coreferential) provided, of course, that they are not *inconsistent*.

---

[1] The best teller of this story is Fodor 1984.

So it is holism that does most of the work in getting radical reference variance. In particular, the culprits are the following two views:

(a) the network of nomological statements specify the descriptive sense of scientific terms in a *holistic* and undifferentiated manner; and
(b) changes in those networks are, always and necessarily, such that they yield *incompatible* descriptions of the same terms.

If these two theses are not independently supported, one cannot get radical reference variance. But there are good reasons to doubt both of them. Even Feyerabend (1965, 259) himself conceded that *not* all theoretical changes lead to changes in meaning and reference. He suggested that the rules (assumptions, postulates) of a theory form a hierarchy where more fundamental rules are presupposed by less fundamental ones. Then, only changes in *fundamental* rules lead to changes in meaning and reference. The fact is that a theory of radical reference variance is untenable *precisely* because it yields an implausible story about how science develops. Here are three reasons for the implausibility of this theory.

First, theories of radical reference variance entail that the referent of a term changes whenever there is even the *slightest* change in the network of descriptions in which the term is embedded. Hence, the theory leaves little, if any, room for *sameness* in reference. It entails that no current theoretical term can have the reference that in the past it was thought to possess, unless in the unlikely event it has retained *in full* its associated network of descriptions *and* nothing more has been added. Second, the theory entails that all scientific disputes about the features of a posited entity were mere equivocations. They could have been avoided had the relevant scientists been prudent enough to point out that they were talking about different entities! Third, it entails that there is no way to assess whether past scientists were right or wrong in their beliefs about the furniture of the world, unless their full web of beliefs has been retained – in which case, they were right!

This attempt to reduce the radical-reference-variance view to absurdity puts the blame on semantic holism and not on the descriptivist theories.[2] With this in mind, let us take a look at the rival causal account of reference and how it fares as a theory of reference-fixing for theoretical terms.

---

[2] It might be thought that the description theories are not good theories of reference anyway. For an effective rebuttal of this claim, see Jackson 1998.

3 THE SWING OF THE PENDULUM II: CAUSAL THEORIES OF REFERENCE

In its original form, this was a theory of the reference of proper names. It was a *causal-historical* theory of proper names in the sense that it rendered the causal-historical chain of events that links current uses of a term to its introduction part of both the reference-fixing and the reference-transmission mechanisms. As introduced by Kripke, the causal-historical approach identifies the reference (denotation) of a name with the individual/entity that this name was used – in an initial act of baptism – to denote. Hence, the reference of a currently used proper name, e.g. 'Aristotle', is fixed by the causal-historical chain which links the current use of the name with an introducing event, during which the name was given to its bearer (*Aristotle*). Descriptions which are, as a matter of fact, associated with the name (e.g. that Aristotle was the founder of the Lyceum) might be false, and yet current name users do refer to the named individual, insofar as their use of the name is part of a transmission chain which goes back to the introducing event. The thrust of the causal theory is that the relation between a word and an object is *direct* – a direct causal link – unmediated by a concept. In particular, the causal theory dispenses with descriptive senses as reference-fixing devices.

This theory is causal in an oblique sense, namely that the reference-transmission mechanism *is* causal. The reference-fixing mechanism is causal too but only in the sense that the introduction of the name is done (and hence its reference is fixed) in the *presence* – that is, the *perceived* presence – of its bearer. It's not as if the presence of a certain child *causes* its name-giver to give him the name 'Aristotle' as opposed to anything else. This part is purely (or almost purely, given several traditions of naming in Greece) conventional. Rather, what happens is that this conventional act of naming picks out its bearer uniquely because there is *causal contact* between the name given and the bearer of the name. So the key thought behind the causal theory of reference is that this causal relation – that underpins the causal contact between the name-giver and the dubbed item – fixes the reference in an unconceptualized way.

When Putnam extended this view to cover the reference of natural-kind and physical-magnitude terms, the idea was that the relevant introducing event (e.g. of the kind term 'gold') involved causal contact with samples of the dubbed stuff. But this is not clearly enough to fix the reference of a natural-kind term, the reason being that, at most, it succeeds in making the term refer to the sample of stuff that is actually present. To succeed in fixing the reference of the term to the natural kind itself (assuming that it is a natural kind) it is also required that in the very act of baptism the introduced term is said to refer to whatever stuff is similar to the one that causally grounds the introduction. So something like the following is needed during the introducing event:

> for all x, x is an F (e.g. an elephant) iff x stands in a specific similarity relation (e.g. sameness of nature) to *this* specific object, picked out by ostension.

As Putnam (1983, 73) put it, 'a term refers (to the object named) if it stands in the right relation (causal continuity in the case of proper names; sameness of "nature" in the case of kinds terms) to these existentially given things'.

Fixing the reference of physical-magnitude terms (e.g. 'electricity') is a variation of the same theme: when confronted with some observable phenomena, it is reasonable to assume that there is a physical magnitude (or entity) which causes them. Then we (or indeed, the first person to notice them) dub this magnitude with a term t and associate this magnitude with the causal production of these phenomena. During this introducing event of the term t as referring to this magnitude, one might typically associate t with a description, i.e. with a causal *story*, of the nature of the posited entity and of the properties in virtue of which it causes its paradigmatic observable effects. This initial description will most likely be incomplete, or even misguided. It may even be wrong: a mistaken account of the nature of this causal agent. But, on the causal theory, one has nonetheless introduced existentially a referent – an entity causally responsible for certain effects to which the term t refers. That is, one has asserted that:

> There is a ϕ [ϕ is causally responsible for certain effects Φ and for all terms t (t refers to ϕ if and only if t picks out the causal agent of Φ)].

The chief attraction of the causal theory is that it disposes of semantic incommensurability.[3] If, for instance, the referent of the term 'electricity' is fixed existentially, all different theories of electricity refer to, and dispute over, the same 'existentially given' magnitude, namely the causal agent of salient electrical phenomena. The causal theory lends credence to the claim that even though past scientists had partially or fully incorrect

---

[3] For definitive work on the issue of incommensurability see Howard Sankey's papers, e.g. 2009.

beliefs about the properties of a causal agent, their investigations were continuous with those of subsequent scientists, since their common aim has been to identify the nature of the same causal agent.

The chief problem with the causal theories of reference is that they make referential success all too easy. If the reference of theoretical terms is fixed purely existentially, then insofar as *there is* a causal agent behind the relevant phenomena, the term is bound to end up referring to it. Hence, there can be no referential failure – even in cases where it is counter-intuitive to expect successful reference. Taken to its letter, the causal theory makes referential success inevitable. For instance, there is no easy way to show within a causal theory that 'phlogiston' was not referring to *oxygen*.

The problem is not with the idea that causation is involved in reference-fixing. It should clearly be part of the story of the referential failure of 'phlogiston' that phlogiston did *not* cause the phenomena for which the term 'phlogiston' was introduced. The problem is that the presence or absence of a cause – in and of itself – cannot be the total determinant of reference. Modes of presentation of the referent should be added.

Actually, the causal theory is consistent with the fact that descriptions may be used to identify a referent. Indeed, there are interesting cases in which the specification of a referent is (or can be) made *only* via a description, e.g. in the case of the introduction of the name 'Neptune' for the newly discovered planet that perturbed the orbit of Uranus, or of the name 'Jack the Ripper' (see Kripke 1980, 79–80, 96). Such cases, it should be noted, involve most occasions in which a theoretical entity is posited as the causal agent of certain phenomena and a name is picked to refer to it. It makes little difference that, as Kripke (1980, 106) put it, the description 'fixes the reference by some contingent marks of the object'. The important thing is that there are clear cases in which not much referential headway can be made without using descriptions.

4   THE PENDULUM'S RESTING PLACE: CAUSAL
DESCRIPTIVISM

If we want to avoid the problems noted above, it seems we have to combine some lessons learned from the problems faced by the standard theories of reference. This would push us towards a causal descriptivist account of reference, one that utilizes insights and resources from causal and descriptivist approaches to reference-fixing.

4.1   The basics

Causal descriptivist accounts are by no means new. The basic *negative* idea behind them is this. The reference-fixing mechanism should not be just a non-conceptual causal relation to the referent; if it were, referential success would be easy to get (and trivialized) insofar as there is indeed some kind of causal contact with the referent. However, the reference-fixing mechanism should not be just a conceptual description-based relation to the referent; if it were, referential stability would be hard to get insofar as the description-based relation to the referent is determined by theories and theories change.

The basic *positive* idea, then, is this. The reference-fixing mechanism should have the following form:

R(x) = x causes phenomena Φ *and* D(x).[4]

Term t refers to x if and only if R(x).

This appears to be a genuine hybrid account. The causal relation with the referent plays an indispensable role in reference-fixing. This is meant to be captured by the claim that ⌜x causes Φ⌝. But the causal relation is not enough. Reference is fixed by means of descriptions of the causal role attributed to the putative referent. This is meant to be captured by D(x). D(x) is a causal description, namely a description of the ways in which the posited referent is supposed to be causally connected with the phenomena Φ that it is taken to cause.

That *each* component is required may be defended as follows. The causal component ⌜x causes Φ⌝ is required because the referents of theoretical terms are introduced as causes of certain phenomena. The descriptive component D(x) is required because the referent (the cause of Φ) should be attributed some properties – those that capture its causal role – if cognitive (as opposed to merely causal) access to it is to be had. That *both* components are required may be seen from the fact that their verdicts may not match: the entity that satisfies D(x) vis-à-vis Φ might not be the entity that does cause Φ. The oxygen/phlogiston case is quite instructive in this respect. It is quite clear that in this case, there is a mismatch between the two components of R(x), if they are taken separately: the cause x of the phenomena Φ of combustion was the presence of oxygen and yet the (phlogiston-based) description D(x) of the cause could not single out

---

[4] A variant of this account is considered but not adopted by Kroon (1985).

oxygen as the cause. The presence of both components of R(x) makes sure of the following: A term t refers to x only if x is the cause of Φ. If causal description D(x) is satisfied by some entity x, but x is not the cause of Φ, then t does not refer to it. Conversely, if something does indeed cause Φ but D(x) is not satisfied by this something, then t does not refer to it.

4.2 Three objections

Before I elaborate further on the basic idea, let me forestall three plausible objections to it. The first is that this hybrid theory is ad hoc, its only motivation being to accommodate counterexamples to the two standard theories of reference. The reply to this objection is that it seems a plausible general condition on reference-fixing that it combine two elements that need not coincide in all cases, namely a causal element and a cognitive one.[5] In ordinary cases of perceptual access to objects – which are the paradigm cases of reference-fixing – the two conditions coincide (or tend to coincide anyway). Perception is a process by means of which we are in causal contact with worldly objects and by means of which we get cognitive access to them and their properties. Even if we do not buy into causal theories of knowledge, it is generally true that the perceptual processes by means of which we get to know things about objects of perception are the very processes by means of which we are causally connected with the objects of perception. There are malign cases, but these need not invalidate the general point. In the case of perceptual contact, R(x) is independently plausible and, as a rule, correct. In the case of fixing the reference of theoretical terms, these two conditions diverge. This is simply because there is no perceptual contact with the referent, though there is causal contact with it (if indeed, there is one). Hence, there must be a way in which cognitive access to the referent is ensured, at least in principle. Otherwise, the entity that does the causing of Φ might not be, even in principle, cognitively accessible. Then, referent-fixing would be a merely existential act: there is something that causes Φ. The cognitive access to this something – and hence the very possibility of knowing what this something is and whether it is indeed the one required by the truth of the theory which posits it – would require some kind of description of it: a description which would offer some *identifying marks*, those, I would add, by means of which it is supposed to play the causal role it does. R(x)

---

[5] This is also suggested by Kroon (1985, 1987, 2009).

then becomes independently plausible: it ensures the cohabitation of the two elements of reference-fixing, i.e. the causal and the cognitive.

The second plausible objection is that the account offered is not *really* a hybrid one – it is, in essence, a *descriptive account* of reference-fixing. The reference, it might be said, is fixed by a causal *description*. A variant of this objection, voiced by Raatikainen (2007), is something we already considered, namely that the original causal theory of reference did allow for descriptions to play a role in reference-fixing but insisted that descriptions may well be insufficient and unnecessary for a unique identification of the referent. On reflection, I think nothing much hangs on this objection. If the causal element in the original causal theory of reference is reduced to the role that causation plays in the *transmission* of reference of a term from those present in the act of baptism to its current users, as Raatikainen implies, this is a very thin causal theory of reference. If, as is more plausible, causation is said to play a more central role in reference-fixing, by linking the name with its referent without the need to interpolate an identifying description, then the objections noted above do suggest that there is more to reference-fixing than causal contact. In any event, if descriptions are allowed to play some role in reference-fixing, it's a semantic issue whether the resulting theory is *really* descriptive.[6] The way causal descriptivism was articulated above does give some genuine role to causation, even if, at the end of the day, it is captured by claims of the form '*x* causes Φ'.

The third plausible objection is certainly the most serious and may be called 'the too little/too much' objection. In essence it is the problem of how rich the description should be, given that it is not the *whole* of the theory (as holistic accounts implausibly required). If the causal description D(*x*) associated with a newly posited entity is very rich, even though it may explain in virtue of what putative mechanisms it is supposed to play its causal role, it will be difficult to ensure that it is satisfied by anything, let alone to guarantee some referential continuity in theory change. The entities posited by successor theories rarely ever take up most of the explanatory-causal structure attributed to abandoned entities. If, on the other hand, the causal description attributed to a newly posited entity is too thin, it is quite likely that it will be

---

[6] Interestingly, the mirror image of this objection is made by Boyd (2010, 224), who claims that causal descriptivist theories are really *causal* theories, since descriptions and their use play a causal role in reference-fixing. Deploying descriptions may well be part of the causal profile of reference-fixing by language users. But the key point is that descriptions offer identifying marks without which the referent cannot be tracked.

multiply satisfied. Hence, there will be no guarantee that a unique refer-ent is picked out.

The obvious answer to this objection is this: $D(x)$ should have the right size! It should be neither too rich, nor too thin. But then how is this pos-sible? It is possible because descriptions associated with theoretical terms play a certain *role* in reference-fixing: they are meant to offer enough iden-tifying markers of an entity (related to its causal role vis-à-vis phenomena Φ) to allow the stable use of the term in certain inductive and explanatory practices; but they are not meant to asphyxiate the putative referent, that is, to leave no room for error, ignorance, or improvement. Theory devel-opment is an ongoing process and, I claim, it is an ongoing concern of the scientists to maximize the chances of referential continuity of theor-etical terms (without, however, making referential continuity inevitable). Hence, care is generally taken to associate with a term enough descrip-tions to pick out its referent uniquely and to ground its putative explana-tory role, but not too many to forestall any referential continuity. This is evidenced by the fact that, as a rule, there is consensus vis-à-vis referential success and failure among scientists, which means that there is consensus as to which were the reference-fixing descriptions and which were not. As a quick example of this, take the much-discussed contrasting cases of 'caloric' and 'electricity'. They were both described as imponderable fluids; they were both taken to have material composition; etc. But it turned out that though these descriptions were reference-fixing for 'caloric', they were not reference-fixing for 'electricity'. When nothing turned out to satisfy them in the case of the cause of heat phenomena, 'caloric' was abandoned as denotationless. But 'electricity' was not abandoned as denotationless, even though the foregoing descriptions were not satisfied; hence, they were *not* reference-fixing descriptions.

Admittedly, there is an element of hindsight in this way of putting things. But it is hindsight available to contemporary practitioners them-selves. Here are two dictionary entries of the terms 'caloric' and 'electri-city' in 1832:

CALORIC, in Chemistry, a modern term introduced into philosophy, to denote that substance, by the influence of which are produced all the phenomena of heat, and which was formerly distinguished by the term igneous fluid, matter of heat, and other analogous demonstrations. (Jamieson 1832, 140)

ELECTRICITY, the name of an unknown natural power which produces a great variety of peculiar and surprising phenomena, the first of which are supposed to have been observed in the mineral substance called amber, whence they have been denominated electrical phenomena, and the laws, hypotheses, experiments, &c

by which they are explained and illustrated, form together the science of electricity. (Jamieson 1832, 255)

Notice the important difference in the way the two referents are described. In the case of 'caloric', it is part of the identifying markers of it that it is a (material) substance. In the case of 'electricity' it is not. Rather, electricity is picked out by descriptions that allow refinement of the putative nature of the power that produces certain phenomena.[7]

More generally, we can follow David Papineau's (1996) suggestion that the descriptions associated with a theoretical term are divided into three parts: the 'yes-part', that is, those that do contribute to the definition of the term; the 'no-part', that is, those that definitely do not contribute to the definition of the term; and, finally, the 'perhaps-part', that is, those descriptions which might (or might not) contribute to the definition of the term. Insofar as the yes-part is rich enough to pick out a referent uniquely, the presence of the perhaps-part (which yields a certain imprecision in the meaning of the term) does not endanger the referential determinacy of the term.[8] What the above discussion has added is that the determination of the yes-part of the definition of a term is by and large brought about by considerations of maximizing the chances of transtheoretic referential stability without making referential success inevitable.

4.3   Some meat on the bones

I take it that an important feature of the causal descriptivist account outlined above is the tracking requirement: a theoretical term t must track

---

[7] This is by no means atypical. Here are the relevant entries in Andrew Ure's *Dictionary* of 1820.

CALORIC. The agent to which the phenomena of heat and combustion are ascribed. This is hypothetically regarded as a fluid, of inappreciable tenuity, whose particles are endowed with indefinite ido-repulsive powers, and which, by their distribution in various proportions among the particles of ponderable matter, modify cohesive attraction, giving birth to the three general forms of gaseous, liquid, and solid. ([1820] 1828, 251)

ELECTRICITY. The phenomena displayed by rubbing a piece of amber, constitute the first physical fact recorded in the history of science. ... From electron, the Greek name of amber, has arisen the science of electricity, which investigates the attractions and repulsions, the emission of light, and explosions, which are produced, not only by the friction of vitreous, resinous, and metallic surfaces, but by the heating, cooling, evaporation, and mutual contact of a vast number of bodies. ([1820] 1828, 401)

Ure, very meaningfully, adds that in the case of caloric there is an alternative account of the causes of heat phenomena, i.e. the vibratory motion of the 'particles of common matter'.

[8] There are certain difficulties with Papineau's account which are discussed by Papineau himself and by Kroon and Nola (2001).

its referent.[9] This means that it must allow the determination of the reference, the acquisition of further information about it, its re-identification, etc. It is obvious that the satisfaction of this requirement presupposes the use of identifying descriptions.

In my (1999) *Scientific Realism*, I developed causal descriptivism on the basis of Enç's (1976) claim that the burden of reference is borne by the *kind-constitutive properties* attributed to the posited entity. Hence, I claimed that the causal descriptions that fix the reference of a term – what I now call $D(x)$ – should capture these properties. This way to proceed can explain referential success and failure. 'Phlogiston' fails to refer because no entity has the kind-constitutive properties attributed to phlogiston. And 'oxygen' succeeds in referring, because there is an entity with oxygen kind-constitutive properties. Since we have no theory-independent access to the kind-constitutive properties of a natural kind, we have to rely on theories and their causal-explanatory descriptions of the entities they posit. There is simply no other way to proceed. But this does not imply that the whole of the theory is implicated in reference-fixing. The kind-constitutive properties are those whose presence makes a set of objects have the same, or sufficiently similar, manifest properties, causal behaviour, and causal powers. Their specification need not involve the whole of the theory.

If we assume that natural kinds have boundaries – based on objective similarities and differences – we can see how causal descriptions succeed in fixing the reference of a term. The one and only entity to which the term refers is the entity which is characterized by the relevant kind-constitutive properties. So, $D(x)$ identifies the referent x of term t in such a way that (a) if no entity satisfies $D(x)$ (i.e. if it is true of no entity), then t does not refer; and (b) if an entity y does not satisfy $D(x)$, y and x cannot play the same causal role. So, on this view, we have a readily available account of referential success and failure.

I still think this idea is essentially correct. But I want to broaden it a bit by claiming that the burden of reference is carried by *any* stable identifying properties provided that they are taken to contribute to the causal role attributed to the posited entity vis-à-vis the phenomena Φ. This loosening is important because what properties are kind constitutive might be a matter that is settled at quite advanced stages of inquiry whereas successful reference can be established at quite early stages. It is hoped that the identifying properties will be part of the kind-constitutive properties of the

---

[9] This tracking requirement is central in Boyd's (2010) accommodationist view of reference of natural-kind terms.

posited entities, but this is not something that has either to be assumed *ab initio* or required in the end. Actually, this kind of latitude allows for reference refinement.

There are two prima facie plausible objections to the claim that stable contingent properties may play a role in reference-fixing. Relying on Kripke's well-known modal argument against descriptivist theories, one may argue that in modal contexts where the referent does not possess these contingent properties, reference may still succeed, given the causal connection. Hence, it might be claimed, there is a danger that the suggested account is descriptivist only in name. Similarly, relying on Kripke's well-known epistemic argument against descriptivist theories, one may argue as follows: if we are wrong about the descriptions which concern putative contingent properties, but still in causal contact with the referent, do we succeed in referring to it? If yes, doesn't this show that it is the causal element that is essential to reference? And, if not, why not?

Though both objections deserve more attention than I offer here, I think meeting them should start by stressing that in the suggested account causation and stable identifying properties (picked out by descriptions) work *in tandem* in reference-fixing. Regarding the first objection, it is precisely an advantage of this account that reference is not fixed by descriptions only. Causal contact serves the purpose of anchoring the descriptions to the referent, thereby rendering these descriptions (at least in the favourable cases) descriptions *of* the referent in the actual world. Conversely (and regarding the second objection), it should be clear that referential success is not merely a matter of causal contact since, as stressed already, it requires that the theoretical terms track their referents, which is achieved by securing identifying marks of the referents, however contingent. To put it succinctly, causal contact anchors the descriptions to the referent; descriptions enable tracking the referent.

To illustrate all this let's discuss briefly the neglected case of chlorine. The term 'chlorine' was first introduced by Humphry Davy in 1812 to refer to what was taken to be dephlogisticated muriatic acid. Carl Wilhelm Scheele had isolated chlorine already in 1774, but being based on the phlogiston theory, he described it as dephlogisticated muriatic acid. He produced it by heating manganese dioxide with hydrochloric acid (called 'muriatic acid'). The idea was that the heating had removed phlogiston from the muriatic acid. There was some controversy as to whether dephlogisticated muriatic acid was an element or a compound. Davy experimented with dephlogisticated muriatic acid, became convinced that it was an element and called it 'chlorine'. Actually, the very term 'chlorine' was introduced to capture the greenish-yellowish colour of the gas (from the Greek word 'chloros', meaning 'green'). But Davy was fully aware of two things. One: the gas he had isolated was the same one that Scheele had isolated.[10] Two: Scheele's description of it (and the general theory on which it was based) was flawed; the key problem was that dephlogisticated muriatic acid (aka *oxymuriatic acid*) did not contain oxygen. It was an elementary substance.[11] Hence, Davy thought that a change of name was appropriate too.[12] In particular, the change of name was meant to suggest not a change of the reference, but a change of how the referent was identified via appropriate properties of it. The detailed story is long and intricate, as it involved a number of experiments in which Davy showed that chlorine did not contain oxygen and that neither did hydrochloric acid. But the bottom line is this. Chlorine is an element (and not a compound); it does not involve oxygen; it has bleaching power (when water is present); it is a constituent of sea salt; it unites with hydrogen (in equal volumes) to produce hydrochloric acid; etc. This looks very much like an associated description of the form $D(x)$ above. It is a description of a causal agent of a certain sort. It is a description by means of associated properties and relations. Davy was open to the fact that these might prove wrong. He was almost certain that they would need refinement, anyway. Ultimately, the reference should be fixed by the kind-constitutive properties of chlorine.[13] But until this is done, there is no reason to think that reference has not been fixed by means of stable characteristic properties. Indeed, it is clear that all subsequent discussion relied on the fixity of the reference

---

[10] 'It is evident from this series of observations, that SCHEELE'S view (though obscured by terms derived from a vague and unfounded general theory) of the nature of the oxymuriatic and muriatic acids, may be considered as an expression of facts' (Davy 1902, 28).

[11] 'I stated a number of facts, which inclined me to believe, that the body improperly called in the modern nomenclature of chemistry, oxymuriatic acid gas, has not as yet been decompounded; but that it is a peculiar substance, elementary as far as our knowledge extends, and analogous in many of its properties to oxygene gas' (Davy 1902, 40).

[12] 'To call a body which is not known to contain oxygene, and which cannot contain muriatic acid, oxymuriatic acid, is contrary to the principles of that nomenclature in which it is adopted; and an alteration of it seems necessary to assist the progress of discussion, and to diffuse just ideas on the subject. If the great discoverer of this substance had signified it by any simple name, it would have been proper to have recurred to it; but, dephlogisticated marine acid is a term which can hardly be adopted in the present advanced era of the science' (Davy 1902, 59).

[13] 'As chemistry improves, many other alterations will be necessary; and it is to be hoped that whenever they take place, they will be made independent of all speculative views, and that new names will be derived from some simple and invariable property, and that mere arbitrary designations will be employed, to signify the class to which compounds or simple bodies belong' (Davy 1902, 61–62).

of 'chlorine' to an element and not a compound.[14] As is well known, the
atomic weight of chlorine, being 35.5 presented an anomaly to Proust's law
of definite proportions. It turned out that samples of chlorine contained
two stable chlorine isotopes (chlorine-35 and chlorine-37 in a ratio of 3:1).

This development is instructive when it comes to reference refinement.
Isotopes are atoms with the same number of protons but different numbers of neutrons in their nucleus. Isotopes of the same element have the
same chemical properties, but different physical ones, e.g. atomic mass.
The very term 'isotope' was introduced to refer to entities (atoms) with
different causal roles vis-à-vis various radioactive phenomena but 'located'
in the same place in the periodic table, that is, having the same atomic
number and chemical properties. It turns out that two kinds of chlorine
play the causal role specified by the description $D(x)$ noted above: they
were different in virtue of properties that differentiated them qua *nuclides*.
A nuclide is an atom with a specific number of protons and neutrons in
its nucleus. Hence, there are two chlorine nuclides and it is arguable that
they capture a finer division into kinds of chlorine. So the initial term
'chlorine' admits of reference refinement which is achieved when the original identifying properties are refined in such a way that there is a better
match between them and the kind-constitutive properties.

I take it that the foregoing account has a number of genuine
advantages.

- It does not make referential continuity inevitable.
- It explains both referential failure and success in a non-trivial way.
- It allows for a further determination of the referent by means of add-
  ed-on descriptions.
- It leaves open the possibility of re-identifying the reference of a term,
  e.g. when it turns out that $x = y$ even though $D(x)$ is different from
  $D'(y)$. This is likely to happen when there is a unifying description that
  brings under its fold both $D(x)$ and $D'(y)$.
- It shows how the associated descriptions $D(x)$ track the referent of the
  term: this happens when they do succeed in identifying the causal agent
  of phenomena Φ.

5   RAMSEY-SENTENCE REALISM AND REFERENCE

There is a view of growing popularity that a Ramsey-sentence approach
to theories will (dis)solve the issue of the reference of theoretical terms.

---

[14] Cf. Turner 1835, 224–226.

Actually, this approach is taken to be part of a broader descriptivist account of reference. Since the readers might not be familiar with the Ramsey-sentence approach, here is a brief summary of it.

To get the Ramsey sentence ^{R}TC of a (finitely axiomatizable) theory TC we conjoin the axioms of TC in a single sentence, replace all theoretical predicates with distinct variables u_i, and then bind these variables by placing an equal number of existential quantifiers $\exists u_i$ in front of the resulting formula. Suppose that the theory TC is represented as TC $(t_1, \ldots, t_n; o_1, \ldots, o_m)$, where TC is a purely logical $m+n$-predicate. The Ramsey sentence ^{R}TC of TC is: $\exists u_1 \exists u_2 \ldots \exists u_n$TC $(u_1, \ldots, u_n; o_1, \ldots, o_m)$. For simplicity let us say that the theoretical terms of TC form an n-tuple $t = <t_1, \ldots, t_n>$, and the o-terms of TC form an m-tuple $o = <o_1, \ldots, o_m>$. Then, ^{R}TC takes the more convenient form: $\exists u$TC(u,o).

Now, as Carnap showed, the theory in the old form (i.e. TC) is *logically equivalent* to the conjunction ^{R}TC & (^{R}TC$\rightarrow$TC). ^{R}TC is the Ramsey sentence of the theory, while the conditional ^{R}TC$\rightarrow$TC – known as the Carnap sentence – asserts that *if* there is a class of entities that satisfy the Ramsey sentence, *then* the theoretical terms of the theory denote the members of this class. Carnap suggested that the Ramsey sentence of the theory captured its factual content (which was expressed in the rich language LT), while the conditional ^{R}TC$\rightarrow$TC captured its analytic content (it is a *meaning postulate*).[15]

Modulo the Carnap sentence, the theory is equivalent to its Ramsey sentence. But no theoretical terms occur in the Ramsey sentence of the theory (by construction). Hence, it might be thought, the problem of the reference of theoretical terms is (dis)solved. Actually, there are two ways to proceed, given the foregoing equivalence. The first is eliminative, the second is vindicatory. According to the eliminative way, the issue of the reference of theoretical terms vanishes. According to the vindicatory way, the issue of the reference of theoretical terms is solved once and for all. Let us take them in reverse order.

Given the equivalence TC iff ^{R}TC & (^{R}TC$\rightarrow$TC), we can think of the theory as implicitly defining its theoretical terms. An implicit definition of a concept F (or a set of concepts F, G, H, …) fixes the meaning of this concept specifying that certain postulates in which it occurs are true (similarly, for a set of concepts). Using standard notation, we may write '#_' for the set of postulates with a blank wherever the *definiendum*, F, occurs; the implicit definition then assumes the form '$F =_{\text{def}} \#F$ is true'. This kind of

---

[15] For more on this, see Psillos 2000, 2008, and Psillos and Christopoulou 2009.

move implies that the entities to which the implicitly defined concepts apply are whatever ones satisfy the postulates (provided, of course, that the postulates are consistent). Actually, *any* system of entities that satisfy the postulates is such that the implicitly defined concepts apply to them. The Ramsey sentence ^{R}TC says that there are classes of entities which are correlated with the observable events in the way the postulates of the theory describe; but it does not say what exactly those entities are – it does not pick out any such class in particular. The Carnap sentence, then, asserts that '*if* anything satisfies the Ramsey sentence of the theory, *then* the entities posited by the theory do'. Hence, the theoretical concepts of the theory are implicitly defined by the Carnap sentence in such a way that they refer to *whatever* entities satisfy the Ramsey sentence. This is clearly an implicit *definition*: it stipulates meanings without defining them explicitly.

It seems the only residual question is this. Are there entities that satisfy the Ramsey sentence of the theory? David Lewis (1970), who made the above approach popular, urged that theoretical terms refer only if there is a *unique* realization of TC, that is, only if there are no multiple classes of entities which stand in relation TC to *o*. If there are more than one such classes, the theoretical terms should be taken to be denotationless. His motivation for this claim was that such a view is the lesser of two evils. In case of non-unique realization, there is no non-arbitrary way to pick one realization. So, Lewis thought, we are forced to accept either that theoretical terms do not name anything, or that they name the elements of one arbitrarily chosen realization. For him, however, 'either of these alternatives concedes too much to the instrumentalist view of a theory as a mere formal abacus' (1970, 432).

If all goes well – if, that is, *uniqueness is achieved* – theoretical terms are implicitly defined *and* their reference is secured: to the unique realization of the theory. That's why I called this kind of view 'vindicatory'. All does *not* go well, however. Since the reason why all does not go well is related to the prospects of the eliminative approach, I should look at it first.

On the eliminative view, the very idea that the Ramsey sentence *replaces* the theory shows that the very issue of the referential success or failure of theoretical terms becomes irrelevant to the appraisal of theory and its claims to truth. All that is required for the truth of the theory is that the Ramsey sentence is satisfied: that there is a realization of it. However, as Lewis was quick to note, the Ramsey sentence might not have an *exact* realization; it might have a near-realization. Let ^{R}TC be the Ramsey sentence of the original theory which has no unique realization, and TC* a theory which is a modification of TC (perhaps, a corrected version of TC). Suppose that TC* has a unique realization R which happens to nearly realize TC. Then, R is a near-realization of ^{R}TC. In the shift from TC to TC* there is, typically, change of reference (since the realizations of TC and TC* are different) unless the realization of TC* is a near-realization of TC too. In this case, it can be plausibly argued that there is referential continuity between TC and TC*. Here is exactly where the eliminative view gets its grip. If the content of the theory is *fully* captured by its Ramsey sentence, then – it is argued – referential failure (or success, for that matter) becomes irrelevant: the realizer of ^{R}TC* is the near-realizer of ^{R}TC and *that's it*. The issue of reference drops out of the picture. Pierre Cruse (2004, 140), who – together with David Papineau (see Cruse and Papineau 2002) – defends this view, goes on to claim that there needn't even be a near-realizer; various realizers might be allowed and the terms of the theory might be taken to 'partially denote' each of them.

The key difference between the vindicatory way and the eliminative way concerns the status of the Carnap sentence ^{R}TC→TC. The eliminative approach takes it to be 'devoid of cognitive content' (Cruse 2004, 146). The vindicatory approach takes it to be cognitively significant. In fact, in Lewis' hands, the Carnap sentence is strengthened by the uniqueness-of-realization requirement. It now reads as follows: '*if* anything uniquely satisfies the Ramsey sentence of the theory, *then* the entities posited by the theory do'. Let's call this the Lewis sentence.[16] Accordingly, the theoretical terms are fully defined (in fact, they can be *explicitly* defined) and their reference is fixed. The original Carnap sentence does not imply uniqueness of realization. It is fully consistent with it that there might be multiple realizers of the theory TC.[17] If there happen to be multiple realizers, the vindicatory approach says the terms of TC fail to refer. If there are multiple realizers, the Carnap sentence does allow that the terms of the theory might refer to them. In fact, as noted above, the Carnap sentence has no means to single out one of the many realizations as the 'correct' one.

---

[16] The Lewis sentence is implied by the Carnap sentence. Lewis adds two more meaning postulates. (1) *If* nothing satisfies the Ramsey sentence of the theory, *then* the entities posited by the theory do not satisfy it. Hence, the terms of the theory are denotationless. Postulate (1) is independent of the Carnap sentence. (2) *If* the Ramsey sentence of the theory is multiple realized, *then* the terms of the theory are denotationless. Postulate (2) is actually inconsistent with the Carnap sentence in that it disallows that the terms of the theory might refer to the entities posited by the theory, if the theory is not uniquely realized by these entities.

[17] Carnap famously thought that there will be mathematical realizers of the theory; see my 2008 for the details.

The real issue, however, is not whether the Carnap sentence has cognitive content or not – though it does: the Carnap sentence poses a certain restriction on the class of models that satisfy the theory; it excludes from it all models in which the Carnap sentence fails. The real issue is with *multiple realization*. It turns out that the Ramsey sentence of the theory is bound to be multiply realized, unless restrictions are imposed on the range of its (second-order) variables. Suppose the Ramsey sentence ^{R}TC of a theory TC is empirically adequate. Subject to certain cardinality constraints, if the Ramsey sentence is empirically adequate, it is true. In other words, there is at least one realization of it and, given plausible assumptions about the universe of discourse of the second-order variables, a multiplicity of them. The proof of this has been given in different versions by several people.[18] Its thrust is this. Since ^{R}TC is consistent, it has a model. Call it M. Take W to be the 'intended' model of TC and assume that the cardinality of M is equal to the cardinality of W. Since ^{R}TC is empirically adequate, the observational submodel of M will be identical to the observational submodel of W. That is, both the theory TC and its Ramsey sentence ^{R}TC will 'save the (same) phenomena'. Now, since M and T have the same cardinality, we can construct a one-to-one correspondence f between the domains of M and W and *define* relations R' in W such that for any theoretical relation R in M, $R(x_1 \ldots x_n)$ iff $R'(fx_1 \ldots fx_n)$. We have induced a structure-preserving mapping of M on to W; hence, M and W are isomorphic and W becomes a model of ^{R}TC.[19]

The Ramsey sentence of a theory may *fail* to be empirically adequate. No one denies that the Ramsey sentence of the theory has empirical content or that it might be empirically false. But the point still remains that if it *is* empirically adequate (if, that is, the structure of observable phenomena is embedded in one of its models), then it is bound to be true. For, as we have seen, given some cardinality constraints, it is guaranteed that there is an interpretation of the variables of ^{R}TC in the theory's intended domain.

This result has a straightforward bearing on both the eliminative and the vindicatory approaches. I take it that it simply undermines the eliminative approach. For it is not just the issue of reference that drops out of the picture. The issue of realism goes by the board too: empirical adequacy collapses to truth. And not just that. There is simply no way to tell even what

---

[18] Winnie 1967, 226–227; Demopoulos 2003, 387; Ketland 2004.

[19] This is a version of the so-called Newman problem that Russell's structuralism faced. For more on this, as well as on Ramsey sentences, see my 2006. Cruse's reaction to this claim is given in his 2005. For an effective rebuttal see Ainsworth 2009.

kind of things the second-order variables of the Ramsey sentence range over. When it comes to the vindicatory approach, things are somewhat more complicated. It is in the spirit of this approach that there should be ways to exclude various interpretations of the Ramsey-sentence variables, aiming to constrain them, at least in principle, to just one. The way to do this, as Lewis emphasized in subsequent work, is to impose causal constraints on the admissible interpretations of the language. That is, it is necessary to move away from purely descriptivist accounts of reference-fixing.

As Lewis (1984) has stressed, the causal constraint on an admissible referential scheme is not just more theory – a causal theory – added to the theory of the world. Rather, it is constraint to which a referential scheme (a realization of the theoretical terms of the theory) must conform, if it is to be the correct one. Perhaps the best way to understand what this constraint might be has been suggested by G. H. Merrill (1980), who argued that it is questionable that realism conceives of the world merely as *a set of individuals*, i.e. as a model-theoretic *universe of discourse*. The world, a realist would say, is a *structured entity*. Its individuals stand in specific relations to one another, or to subsets of individuals. In particular, a realist would assert two things: (1) the constituents of the world (individuals and properties/relations) are independent of any particular representation we have of it; and (2) the world is *already* structured, *independently* of the language. Under these assumptions, it is easy to see how the Ramsey sentence of the theory might fail to be true, even though it might be empirically adequate. Among all those relations-in-extension which satisfy the Ramsey sentence, only those which express *real relations* should be considered. But specifying which relations are real requires knowing something *beyond* the Ramsey sentence of the theory, namely which extensions are 'natural', i.e. which subsets of the power set of the domain of discourse correspond to natural properties and relations.

The view that the world is a *structured domain* places a *constraint* on the interpretation of a language, namely that referents must be *eligible*. Lewis' eligibility constraint is *prior* to the interpretation of a language and such that an interpretation must satisfy it in order to be intended. Lewis does not suggest an eligibility *theory* which is open to reinterpretations. He offers a constraint and suggests that we have to turn to physics and its 'inegalitarianism' in order to find the elite things and classes of things that constitute the joints of the world. Then he argues that an interpretation is unintended – and it would be disqualified – if it employs gerrymandered referents, i.e. putative referents that do not belong to the objective structure of the world. It follows that if 'we limit ourselves to the eligible interpretations, the ones that respect the objective joints in nature, there is no longer any guarantee that (almost) any world can satisfy (almost) any theory' (1984, 227). There is now a way to eliminate many of the possible realizers of the Ramsey sentence: only those that are eligible remain, where the eligibility of a realizer is a function of the eligibility of the classes over which the variables of the Ramsey sentence vary.

6 CONCLUSION

From a realist point of view, an important constraint on the theory of reference – if not the *raison d'être* of a theory of reference – is the accommodation of linguistic practices to the causal structure of the world. Given that this accommodation is, as a matter of fact, an historical process, it is vital for a realist view of science that the theory of reference allows for convergence in the scientific image of the world. Theoretical terms are necessary for scientists to talk about posits not given in experience; but they should be taken to have transtheoretical robustness. Unless continuity in theory change and convergence are established, past failures of scientific theories will act as defeaters of the view that current science is on the right track. The causal descriptivist approach to the reference of theoretical terms is fit to strike a balance between, on the one hand, the theory dependence of theoretical terms and, on the other hand, viewing theoretical terms as relevantly transtheoretical. This is reason enough to be worthy of development.

REFERENCES

Ainsworth, Peter (2009) Newman's objection. *British Journal for the Philosophy of Science* 60: 135–171.
Boyd, Richard (2010) Realism, natural kinds, and philosophic methods. In H. Beebee and N. Sabbarton-Leary (eds.) *The Semantics and Metaphysics of Natural Kinds*. London: Routledge.
Carnap, Rudolf (1936) Testability and meaning. *Philosophy of Science* 3: 419–471.
Cruse, Pierre (2004) Scientific realism, Ramsey sentences and the reference of theoretical terms. *International Studies in the Philosophy of Science* 18: 133–149.
    (2005) Ramsey-sentences, structural realism and trivial realisation. *Studies in History and Philosophy of Science* 36: 557–576.
Cruse, Pierre and Papineau, David (2002) Scientific realism without reference. In M. Marsonet (ed.) *The Problem of Realism*. Aldershot: Ashgate, pp. 174–189.
Davy, Humphry (1902) *The Elementary Nature of Chlorine: Papers by Humphry Davy (1809–1818)*. Edinburgh: Alembic Club.

Demopoulos, William (2003) On the rational reconstruction of our theoretical knowledge. *British Journal for the Philosophy of Science* 54: 371–403.

Enç, Berent (1976) Reference of theoretical terms. *Noûs* 10: 261–282.

Feyerabend, Paul (1965) Problems of empiricism. In R. Colodny (ed.) *Beyond the Edge of Certainty*. Englewood Cliffs, NJ: Prentice-Hall.

Fodor, Jerry (1984) Observation reconsidered. *Philosophy of Science* 51: 23–43.

Jackson, Frank (1998) Reference and description revisited. *Philosophical Perspectives* 12: 201–218.

Jamieson, Alexander (1832) *A Dictionary of Mechanical Science*. London: Henry Fisher, Son & Co.

Ketland, Jeff (2004) Empirical adequacy and Ramsification. *British Journal for the Philosophy of Science* 55: 287–300.

Kripke, Saul (1980) *Naming and Necessity*. Oxford: Blackwell. Originally published in D. Davidson and G. Harman (eds.) *Semantics of Natural Language*. Dordrecht: Reidel, 1972, pp. 253–355.

Kroon, Fred (1985) Theoretical terms and the causal view of reference. *Australasian Journal of Philosophy* 63: 142–166.

(1987) Causal descriptivism. *Australasian Journal of Philosophy* 65: 1–17.

(2009) Names, plans, and descriptions. In D. Braddon-Mitchell and R. Nola (eds.) *Conceptual Analysis and Philosophical Naturalism*. Cambridge, MA: MIT Press, pp. 139–158.

Kroon, Fred and Nola, Robert (2001) Ramsification, reference fixing and incommensurability. In P. Hoyningen-Huene and H. Sankey (eds.) *Incommensurability and Related Matters*. Dordrecht: Kluwer, pp. 91–121.

Kuhn, T. S. ([1962] 1970) *The Structure of Scientific Revolutions*, 2nd enlarged edn. University of Chicago Press.

Lewis, D. (1970) How to define theoretical terms. *Journal of Philosophy* 67: 427–446.

Lewis, D. (1984) Putnam's paradox. *Australasian Journal of Philosophy* 62: 221–236.

Merrill, G. H. (1980) The model-theoretic argument against realism. *Philosophy of Science* 47: 69–81.

Papineau, David (1996) Theory-dependent terms. *Philosophy of Science* 63: 1–20.

Psillos, Stathis (1999) *Scientific Realism: How Science Tracks Truth*. London: Routledge.

(2000) Carnap's 'Theoretical Concepts in Science'. *Studies in History and Philosophy of Science,* Part A, 3: 151–172.

(2006) Ramsey's Ramsey-sentences. In M. C. Galavotti (ed.) *Cambridge and Vienna: Frank P. Ramsey and the Vienna Circle*. Dordrecht: Springer, pp. 67–90.

(2008) Carnap and incommensurability. *Philosophical Inquiry* 30: 135–156.

Psillos, Stathis and Christopoulou, Demetra (2009) The a priori: Between conventions and implicit definitions. In N. Kompa, C. Nimtz, and C. Suhm (eds.) *The A Priori and Its Role in Philosophy*. Paderborn: Mentis, pp. 205–220.

Putnam, Hilary (1962) What theories are not. In E. Nagel, P. Suppes, and A. Tarski (eds.) *Logic, Methodology, and Philosophy of Science.* Palo Alto, CA: Stanford University Press, pp. 240–251.

(1983) *Realism and Reason*, vol. III of *Philosophical Papers.* Cambridge University Press.

Raatikainen, Panu (2007) Theories of reference and the philosophy of science. Paper presented at EPSA07: 1st Conference of the European Philosophy of Science Association Madrid, 15–17 November.

Sankey, Howard (2009) Scientific realism and the semantic incommensurability thesis. *Studies in History and Philosophy of Science* 40: 196–202.

Schlick, Moritz ([1932] 1979) Form and content: An introduction to philosophical thinking. In Moritz Schlick, *Philosophical Papers*, vol. II. Dordrecht: Reidel, pp. 285–369.

Turner, Edward (1835) *Elements of Chemistry.* Philadelphia: Desilver Thomas & Co.

Ure, Andrew ([1820] 1828) *Dictionary of Chemistry*, 3rd edn. London: Thomas Tegg.

Winnie, John (1967) The implicit definition of theoretical terms. In *British Journal for the Philosophy of Science* 18: 223–229.

(1970) Theoretical analyticity. In R. Cohen and M. Wartofsky (eds.) *Boston Studies in the Philosophy of Science*, vol. VIII. Dordrecht: Reidel, pp. 289–305.

CHAPTER II

Scientific representation, denotation, and explanatory power

Demetris Portides

I INTRODUCTION

In the last few decades, various debates in philosophy of science over how scientific theories relate (referentially or not) to the world has strengthened the view that if this quest is to lead to any fruitful results then three intertwined concepts must be understood: model, representation, idealization. That these concepts are intertwined is evident. A model is meant to represent something, whether an actual or an ideal state, whether a physical or an ideal system. For instance, a model of a building is a representation of an actual (or actualizable) building. Moreover, a model represents a physical system in an abstract and idealized way. That is, a model of a building is not meant as an exact replica but as an idealized and abstract representation of an actual building because, for instance, it represents only certain features of the actual system, e.g. some spatial relations, and ignores others, e.g. the plumbing system. In science one encounters several kinds of models, such as iconic or scale models, analogical models, and mathematical models. In this chapter, my discussion is restricted to mathematical models, to which I shall refer as scientific models. Representation seems to be a primary function of scientific models, and idealization (and abstraction) seems to be the steering thought process by which this function is achieved. By highlighting this point I mean to suggest that a better understanding of 'scientific representation' could be achieved if we examine it in relation to 'scientific models' and to 'idealization'.

It is true that in science various means of representation are used. We say that a diagram of an electric circuit represents its target; that a graph of velocity plotted against time represents the acceleration of a body; that a material construction of double helical structure represents a DNA molecule; that a Feynman diagram represents a neutron decaying to a proton, an electron, and an anti-neutrino; and that a mathematical model represents the behaviour of a mass-spring system. Whether our scientific

representations are diagrammatic, graphical, material, model based, or
other, they are important aspects of scientific inquiry; they enhance our
understanding of the workings of physical systems and often enhance
our understanding of abstract theoretical propositions. Admittedly, if one
aims for a general theory of scientific representation then the latter must
account for all these kinds of representational vehicles. Such an aim, however, goes beyond the scope of this work. Here, I shall confine my analysis
to how scientific models represent physical systems. It is my belief that
scientific models are the most important form of scientific representation, not only because they are widely used, but most importantly because
they are the means by which theories are applied to phenomena. Hence,
the representation relation between model and target system can teach us
about the relation between scientific theories and the world.

One can categorize existing accounts of scientific representation into
two types. Firstly, there are accounts that attempt to reduce representation
to other relations. The semantic view of scientific theories, for instance,
relies on the construal of the representation relation as mathematical mapping. According to the semantic view, the representational relata are mathematical structures, one belonging to the side of theory (theoretical model)
and the other belonging to the side of experiment (data model). Science
represents the world by developing theoretical models which are, according to da Costa and French (2003), partially isomorphic to respective data
models or, according to van Fraassen (1980), their empirical substructure
is isomorphic to respective data models. Data models are constructed by
using the raw data gathered from experimental measurements and refined
and structured according to theoretical dictums. Secondly, there are non-
reductive accounts that try to point out the general features of the representational relation or try to analyse its function in scientific modelling.
The second category could be divided into two: denotative and non-denotative accounts of representation. The inferential account, advocated by
Suárez (2004), is an example of a non-denotative account. According to
this account, for X to be a representation it is necessary that the practices of modelling a phenomenon direct the practitioner to a consideration of the target in a non-arbitrary manner, and the resulting model
allows the user to draw inferences about the target. It, therefore, attributes
two general characteristics to representation. Firstly, the representational
vehicle determines the representational target not by mere stipulation but
through the particular scientific practices. Secondly, the representational
vehicle must be such as to allow the user to draw valid inferences about
the target. Finally, there are denotative accounts that interpret the concept

of representation as strongly linked to the function of denotation, e.g. Hughes (1997) and Elgin (2009). Such accounts construe the representational function of models as denoting component parts and activities of target systems.

In this chapter I shall not deal with non-denotative accounts of representation whether reductive or non-reductive. I focus exclusively on denotative accounts and attempt to develop a denotative account of scientific representation that ties the representational function of scientific models to their explanatory power. In Sections 2 and 3, I explain why existing denotative accounts are plagued with some weaknesses that prevent them from accurately capturing important elements of scientific modelling. In Section 4, I argue that denotative accounts must make use of the notions of 'mechanism' and 'explanatory power' if they are to overcome those weaknesses and do justice to how scientific models represent their target systems.

2   DENOTATION AT THE CORE

Nelson Goodman (1976) argued convincingly that to understand the representation relation, one should dissociate it from the relation of resemblance because of the logical differences between the two concepts. Resemblance is reflexive and symmetric whereas representation is not. Furthermore, Goodman suggested that what lies at the core of representation is denotation.[1] According to Goodman, if X represents Y then X must denote Y. He, as well as other philosophers who opt for an analysis of representation based on denotation (e.g. Elgin 2009; Hughes 1997), realized that by relying only on this idea we run into problems. There are two reasons for this. The first is that X could be considered to be a representation of Y by mere stipulation (call this the stipulation problem). When we decide, in a certain context, to draw crosses on the board to represent cars and draw circles to represent buses, the crosses denote cars and the circles denote buses in that particular context. The second reason is that in many cases X represents Y but Y does not exist and thus there is nothing that X denotes (call this the non-existence problem). If a picture represents something that does not exist, for example a unicorn, it does not make much

[1] For Goodman, questions such as 'what a painting denotes', 'what it expresses', 'what it exemplifies' relate to different varieties of reference. He thus understands denotation to be one particular variety of reference. In the context of scientific representation, which is the focus of my discussion, such distinctions between varieties of referential relations seem to me to be unnecessary, hence I use denotation and reference interchangeably.

sense to say that it also denotes it. Both of these problems are important when one is addressing questions about scientific representation. Firstly, we do not think that we do justice to scientific practices by thinking that scientific models represent by mere stipulation; we think that there are appreciable differences between scientific representations and ad hoc representations such as the crosses and circles on the board. Secondly, some of our scientific models represent what we often label ideal systems or ideal states of affairs, and if such systems do not exist in the actual world then it would also not make much sense to claim that our models denote such systems. In addition, science often models some domain of the world by employing concepts that in retrospect we can firmly claim do not refer to the world, e.g. aether models, or phlogiston models. So if such models represent the world they surely do not do this in virtue of their terms denoting existing entities in the world.

Goodman suggests a way to overcome the non-existence problem. He distinguishes between 'being a representation of Y' and 'being a Y-representation'. If X is a representation of Y then X must denote Y, hence Y must exist. But if Y does not exist we can still think of having a Y-representation. The latter concerns only the kind of representation at hand. More precisely, following Goodman the following three claims should be kept distinct. If X is a representation of target Y, then X denotes Y. If X is a symbol for Y but Y does not exist, then X is a Y-representation. Finally, there are cases where X represents as Y a target Z; in such cases X is a Y-representation denoting Z. The first case, according to Goodman (1976, 28), concerns what a symbol denotes; the second, only the classification of the symbol; and the third, both the denotation and classification of the symbol. Although this idea still needs to be augmented to escape the stipulation problem, it seems a useful tool for exploring the nature of scientific representation.

Goodman uses the above distinctions to make sense of representation in domains other than science. In thinking about representation by means of paintings his idea avoids the non-existence problem. Some of the examples he uses help to clarify the point. We may think of a painting representing Churchill as denoting him. We may think of a painting of a unicorn as being a unicorn-painting. And we may think of a painting representing Churchill as a bulldog as being a bulldog-painting denoting Churchill. More generally, if a painting is a representation of Y then Y is denoted by the painting. And a painting is a Y-representation, depending on what kind of a painting it is. Finally, a painting represents Y as

X if the painting is an *X*-representation of *Y*. All this seems to work well for representation in art; does it also assist us in understanding scientific representation?

To address this question we need to note an important characteristic of scientific models. It is widely accepted that models are constructed by imposing idealizing assumptions that simplify the characteristics of their target systems. Of course, there are different degrees of simplification. Often a model of a high degree of idealization may be used to construct another model of the same target but of a lower degree of idealization. This is accomplished by methods of de-idealization of the initial model. Nevertheless, neither model presents a complete description of the complexities of its target system. Even those models that involve a high degree of detail of their targets are not free of idealizations of one kind or another. This is easy to see by means of a familiar example. In classical mechanics the pendulum system is modelled by the linear harmonic oscillator (LHO). The latter is a second-order differential equation of position with respect to time, which is set up by assuming that the target consists of a point-mass bob supported by a massless inextensible cord of length *l*, performing infinitesimal oscillations of angle θ about an equilibrium point in the absence of a medium. The equation of motion for such a model is:

$$\ddot{\theta} + (g/l)\,\theta = 0 \tag{1}$$

The idealizations involved relate to each of the four underlying assumptions, that the cord is massless and inextensible, that the oscillations are infinitesimal, that the bob is a point mass, and that a medium is absent. When these assumptions are relaxed, then a number of factors must be reintroduced into the equation of motion if the model is to become a reasonably good candidate for predicting the earth's acceleration due to gravity and for explaining the behaviour of an actual pendulum in the lab. The following are the factors accounted for by Nelson and Olsson (1986) in their detailed work on the physics of the pendulum: (1) finite amplitude, (2) finite radius of the bob, (3) mass of the ring, (4) mass of the cap, (5) mass of the cap screw, (6) mass of the wire, (7) flexibility of the wire, (8) rotation of the bob, (9) double pendulum of the support ring and pendulum bob, (10) buoyancy, (11) linear damping for the wire, (12) quadratic damping for the bob, (13) decay of finite amplitude, (14) added mass to the bob's inertia, (15) stretching of the wire, and (16) motion of the support. If the influence of each of these factors is introduced into the

equation of motion of the LHO, the result is an equation of motion that
has the following structure:

$$\ddot{\theta} + (g/l)\ \theta + f_1 + \ldots + f_i + \ldots + f_{16} = 0 \qquad (2)$$

where the f_is are the contributions of each of the above factors. Let me
refer to the model expressed through Equation (2) as the detailed oscillator (DO). There is an obvious difference in degree of idealization between
the LHO and the DO models. The DO is not as idealized as the LHO.
Nevertheless, it contains its own idealizations, despite being one of the
few scientific models that captures most features of its target (experience
shows that its closeness to its target is a limit-case scenario for scientific
modelling in general). For instance, although in the DO the bob is not
assumed to be a point mass, it is nevertheless assumed to be a perfect
sphere of uniform mass distribution. Furthermore, although the angle of
oscillation is not assumed to be infinitesimally small, the oscillations are
assumed to be finitely small but operating in a uniform gravitational field.
It is trivially true that the LHO involves a greater degree of idealization
than the DO, and thus the DO is closer than the LHO to the target system. However, we can also claim that both models involve idealizations;
none of them is an exact replica of the target physical system.[2]

Let us return to the question whether Goodman's suggestion helps for
understanding scientific representation. How should we interpret the
claim that the LHO represents the pendulum? Following Goodman, the
claim can be understood in three ways. The first is that the LHO is a
representation of the pendulum, in which case the constituent parts of the
model denote aspects of the pendulum. However, the constituent parts
of the model are about such things as 'point mass', 'infinitesimal oscillations', etc. The existence of such things is impossible to substantiate.
Similarly, if the DO is thought of as a representation of the pendulum it
would not make much sense to claim that such things as a perfect sphere,
uniform mass distribution, and uniform gravitational field are denoted.
Therefore, if we think of scientific models as *representations of* their target
systems we cannot opt for a view of scientific representation that ties it to
denotation.

The second way is that the model is an LHO-representation (and similarly the other model is a DO-representation). In other words, all the

---

[2] A more detailed argument to the effect that 'de-idealized' models, such as the DO, still involve a
certain degree of idealization can be found in Morrison 1999. In the same paper Morrison explains
in more detail some of the idealizations and approximations involved in the DO.

model does is to tell us what kind of a representation it is. This may be an acceptable way to make sense of what unicorn-pictures are and how they relate to their subject matter, but it is not for scientific representations, for many obvious reasons. One reason is that there are significant differences between the LHO and the DO that we would like to explain; for instance, we consider the DO to be a more accurate model than the LHO (call this the comparison problem). But we cannot and do not need to make a similar claim in comparing two different pictures of a unicorn. If our idea of scientific representation is simply that of classification it does not take us far in comparing different representations and distinguishing which one is the better one.

The third way is that the model is an LHO-representation of the pendulum, i.e. the model represents the pendulum as an LHO. This means that elements of the model denote aspects of the pendulum but as a whole the model also tells us what kind of a representation it is. This avoids the non-existence problem, since some elements of the model denote physical entities or processes, albeit within an overall context of representation that involves idealized concepts and idealized states. As I shall argue later, with proper refinements this view allows us to make the required comparison between the LHO and the DO. The first model represents the pendulum as an LHO and the second model represents the pendulum as a DO. Where the phrase 'the model represents the pendulum as an LHO' is shorthand for 'representing the pendulum as if it consists of a point-mass bob supported by a massless inextensible cord of length l, performing infinitesimal oscillations of angle θ about an equilibrium point in the absence of a medium'. Likewise the phrase 'the model represents the pendulum as a DO' is shorthand for 'representing the pendulum as if it consists of a perfect sphere of uniform mass distribution, performing finitely small oscillations in a uniform gravitational field, and so on'. Since all models involve some degree of idealization of their respective targets it is reasonable to suppose that they function as X-representations of their targets. Despite the fact that if we accept Goodman's suggestion for understanding scientific representation in this manner it does not trivially follow that representing the pendulum as a DO is more accurate than representing it as an LHO, it is still a starting point by which to deal with the comparison problem.

Hughes (1997) recognizes that if representation is to be understood in terms of denotation then representation-as is essential to how scientific models function. Elgin (2009) takes up this point and tries to adjust it to overcome the stipulation problem that representation-as is faced with. In her attempt, she makes use of another of Goodman's notions: exemplification. Exemplification, according to Goodman, is both reference to and possession of a property (or instantiation of a property, in Elgin's vocabulary). If X exemplifies Y, then X refers to Y and possesses Y. Consider Goodman's example of a tailor's booklet of small swatches of cloth. The swatches are meant to exemplify certain properties. As Goodman (1976, 53) explains, the swatch gives a sample of colour, weave, texture, and pattern. Of course, the actual material to be used by the tailor has other properties too, which are not exemplified by the swatch. The swatch exemplifies only those properties that it refers to and is an instance of. Although it is not quite obvious how Goodman's notion of exemplification can be used to gain insight even for artistic forms of representation, I shall not concern myself with this issue here.

3   GENERAL VS. SPECIFIC REPRESENTATION

What is important for my purposes is that Elgin adopts this idea and claims that scientific models exemplify properties of their targets and thus afford epistemic access to them. More precisely her claim is that being a Y-representation, a model '... exemplifies certain properties and imputes those properties ... to its target' (2009, 85). This is how Elgin explains her view:

An ideal gas representation is a fiction characterizing a putative gas that would exactly satisfy the ideal gas law. Such a gas is composed of perfectly elastic spherical particles of negligible volume and exhibiting no intermolecular attractive forces. It exemplifies these properties and their consequences, and thereby shows how such a gas would behave. ... The model is a representation – a denoting symbol that has an ostensible subject and portrays its ostensible subject in such a way that certain features are exemplified. It represents its target as exhibiting those features. So to represent helium as an ideal gas is to represent it as composed of molecules having the features exemplified in the ideal gas model – elasticity, mutual indifference, the proportionality of pressure, temperature, and volume captured in the ideal gas law. (2009, 84)

I could not agree more with Elgin that scientific models afford epistemic access to their target's behaviour, but is it because they exemplify their properties? It may be the case that by invoking the notion of exemplification, the stipulation problem is overcome. But how adequate is this notion in accounting for how mathematical models represent? Exemplification may in some instances be a characteristic of iconic representation, in which case certain predicates are represented via one or more of their instances. Let's return to Goodman's booklet of cloth swatches. The predicate 'is red' or the property of being red (and similarly for other colours or textures) is represented in the booklet by one of its instances or, to use a language closer to Goodman's, by a vehicle of representation that possesses the property of being red. Suppose, however, that the vehicle of representation is language and we represent the colour predicates found in the tailor's booklet by means of a list of wavelength values, e.g. red is represented by wavelength value a, blue by wavelength value b, and so on. Can we claim that the values are instances of the respective property? I think not. When we move away from iconic forms of representation, the notion of 'exemplification' does not shed light on the characteristics of 'representation'. Scientific models (i.e. the mathematical models I am concerned with in this chapter) are not iconic forms of representation, they are expressed by means of equations (in most cases differential equations) and as such not only does it make little sense to say that they instantiate the properties of their subject matter in some way or other, but in fact it obscures their nature and also distorts the functions of their constitutive elements. Elasticity, for instance, is not a property instantiated by the ideal gas model but a condition that has to be met if the relation between pressure, temperature, and volume that the model dictates is to hold for a gas. This goes to say, that models are far more complex entities than what the exemplification view suggests.

Some clarifications of what I have called the comparison problem are needed. Let me assume for the sake of argument that it is tenable to think of a model as exemplifying certain putative features. This view is still faced with the comparison problem. That is, identifying representation with exemplification does not allow us to discriminate between competing representations. Let's return to the LHO and DO examples to clarify this. If we adopt Elgin's view, it follows that the LHO is an entity that exemplifies the feature of a linear restoring force acting on a massless body. It also follows that the DO is an entity that exemplifies a linear restoring force acting on a body, combined with various forces due to the medium, various frictional forces due to the supporting apparatus, etc. In each of the two cases different features are exemplified. Whether those features exemplified by the DO are a better way to represent the pendulum or not, is considered to be a separate question which is dissociated from the notion of representation. However, the question whether a certain model is a better representation of a target system as compared with another model is not the same as the question whether the model is an acceptable representation in the first place.

Elgin attends only to the latter question and seems to suggest that the question of whether a representation X fits its target is independent of what makes X a representation in the first place. Although I would not entirely agree with this, here I only wish to point out that if we opt for a denotative account of representation then we cannot tackle the comparison problem in the same manner as we would in addressing the question of what makes a model an acceptable representation, by treating it as simply a matter of fit to experimental data. This is because two competing representations could have experimentally (or mathematically) indistinguishable predictions, as demonstrated with the following simple example. Let's assume that the competing models are the LHO and model X, where model X involves only two of the influencing factors involved in the DO. Furthermore, assume the physically possible situation where each of these two factors exerts such influence on the oscillations of the pendulum that the two exactly cancel out each other's influence (i.e. one impedes the oscillation and the other assists it in equal amounts). If this were the case, then there would be no predictive difference between the LHO and model X. With respect to how well they fit experimental data the two models would be on a par but, following the exemplification view, the two models exemplify different features and, hence, their representational capacity must differ. But the exemplification view does not give us grounds to discriminate the two representations. I think that it is not sufficient for denotative accounts of representation to rely exclusively on fit to experimental data to address the comparison problem.

What the exemplification view overlooks is that the representational function of models is strongly coupled with the idealizational processes that enter into the model's construction. In fact, idealization *determines* what the models represent. A highly idealized model, such as the LHO, represents the general features of its target. A less-idealized model, such as the DO or the two-factor model above, represents the general features but in addition it also either represents some of the specific features of its target or it represents specific manifestations of the general features. If idealization was not involved in modelling then we would not be thinking of models as putative representations but probably as exact replica of their target systems. Of course, in such an event, talk about the notion of scientific representation would be unnecessary. Hence I take the use of idealization to be an intrinsic characteristic of scientific representation, and as such it brings with it the demand that we distinguish between general and specific representations. The distinction between general and specific representations, which is the primary facet of what I have called the comparison problem, should be reflected in our account of scientific representation. As I have shown above, the exemplification view does not exhibit this characteristic of scientific representation.[3]

Hughes' DDI (denotation, demonstration, and interpretation) account of scientific representation is also inadequate to deal with the general/specific distinction. The DDI account claims scientific models, i.e. the vehicles of representation, involve three components. Firstly, the elements of the models denote elements of their target physical system. Secondly, the models possess an internal dynamic that allows us to demonstrate certain hypothetical conclusions about the target systems. Finally, these theoretical conclusions are interpreted to yield predictions about the behaviour of the target system. So in the DDI account denotation coupled with interpretation is the alternative for exemplification. Despite this, just like the exemplification account, the DDI account attributes to models that represent their targets in a general manner (such as the LHO) and to models that represent the specific features of their target (such as the DO) the same representational features and qualities. The representational accuracy of the two kinds of models is an issue supposedly dealt with independently of their representational function by assessing the accuracy of their predictions. In other words, just as for the exemplification view, an implication of the DDI model of representation is that how well X fits its target is independent of what makes X a representation. But as I have argued we cannot rely on this assumption to overcome the comparison problem.[4]

---

[3] I do not mean to suggest that all idealizations can be ranked in terms of their closeness to the actual characteristics of their targets. This may be true of Galilean idealization (see McMullin 1985) that characterizes the LHO and DO example, above. But, as some authors have argued (e.g. Morrison 1997 and 1999), the process of idealization is far more complex in modern science. For example, often different idealizations lead to the development of not necessarily compatible models, each of which is used to explain and predict a different behaviour of the same physical system. In such cases it is not possible to compare the two or more models and infer which of them is more accurate of the target characteristics. Actually comparing them in terms of their accuracy does not yield any fruitful conclusions; they are simply used for different purposes and there may not be an independent criterion for comparing them. However, it is still possible to determine whether a model's underlying idealizations imply representation of the general or the specific features. If a model is built with idealizing assumptions that totally omit the physical constraints of the target then it can only capture the general features of its target.

[4] I think that the difficulties in dealing with the general/specific distinction present a challenge to accounts of representation that attempt to relate the latter concept to denotation. Other non-denotative accounts of representation such as the one commonly understood to be congruent with the semantic view of scientific theories, in which the representation relation is identified with isomorphism (or partial isomorphism) of structure, have an advantage in regard to this issue. This is so because the degree of representational accuracy of a model, for such accounts, is supposed to be

Earlier I argued in support of Elgin's and Hughes' suggestion of construing scientific representation as representation-as. In particular, I defended the view that if we set denotation at the core of representation then we must construe scientific models as being X-representations of their target systems. I explained why this view overcomes the non-existence problem and raised doubts about whether the exemplification and the DDI accounts can overcome the stipulation problem without running into the challenge of distinguishing between general and specific representations. Both the exemplification and DDI accounts suggest a particular understanding of 'representing-as', which is that all models are indistinguishable with respect to how they represent the features of their targets. Any difference in the representational capacity of different models is supposedly established by comparing their predictions to experimental results. However, in many cases our general intuitions about models lead us to believe that a certain model is a more accurate representation of its target than its competitors, e.g. that there is a certain closeness between the DO and the actual pendulum that the LHO does not have. Experimental results by themselves are not only insufficient to establish this for all cases, but often they are not necessary. In the next section, I suggest a refinement of the denotative account of representation to accommodate the fact that some models represent physical systems by picking out general features and other models also pick out more specific features. The two forms of representation manifest some obvious differences, and developing an account that does not obscure these differences can in my view rescue the denotative approach to representation from the comparison problem and also more accurately capture our intuitions about scientific models.

4   REPRESENTATION AND EXPLANATORY POWER

Although there is no consensus among philosophers of science on what scientific models are, it is widely accepted that some models represent only the general features of their targets whereas others also represent the specific features. Most models that represent the general features are theory driven, i.e. they are direct conceptual descendants of theory. Theory-driven models involve highly abstract and idealized descriptions that satisfy the theoretical postulates. If they represent physical systems they

---

measured in terms of the accuracy of the mapping of elements and relations of a scientific model onto respective elements and relations of a data structure.

do so by picking out the general features of their targets without accounting for many factors that physically constrain the behaviour of these systems. Examples of such theory-driven models are the LHO, the classical wave equation, and the quantum-mechanical harmonic oscillator well. Although it is mathematically possible to develop an indefinite number of such models using the basic postulates of theory, as Cartwright (1999) correctly pointed out, in actual scientific practice only a handful of them are useful. It is possible, for example, to use Newton's laws of motion and the law of gravitation to write down any equation of motion one wishes. Most of these equations are, of course, mathematically intractable but this is not the point here. The overwhelming majority of such equations are not useful for representational purposes in actual science, and more often than not scientists resort to other modelling techniques to carry out the representational task. These techniques usually lead either to outright phenomenological models or to the supplementation of theory-driven models with phenomenological elements (e.g. the DO). Both of the latter are constructed by the use of a conceptual apparatus that is not a direct descendant of theory. This conceptual apparatus is used to account for the particular physical constraints that affect the behaviour of respective physical systems. Such models represent the specific features of physical systems and thus purport to represent physically possible states of their targets.[5]

Such models account for the physical constraints of their targets by providing descriptions of putative mechanisms and of the latter's constitutive parts at work in physical systems. The primary function of these mechanisms is to explain the observed behaviour of the target systems. This is how Morrison (1999) puts this point: 'the reason that [phenomenological] models are explanatory is that in representing [physical] systems they exhibit certain kinds of structural dependencies. The model shows us how particular bits of the system are integrated and fit together in such a way that the system's behaviour can be explained' (63). What Morrison calls integration and fitting together of particular bits of systems I call physical mechanisms. Without significantly deviating from Morrison's view we can make sense of the representational capacity of phenomenological models

---

[5] There are plenty of examples of phenomenological models in modern science, e.g. those of the nuclear structure, such as the liquid-drop model or the unified model. Other examples of phenomenological models from modern physics that have been analysed by several authors, and can be used to make the case for the argument that follows, are: the BCS (Bardeen–Cooper–Schrieffer) model of superconductivity (see Cartwright 1999), the MIT bag model (see Hartmann 1999), the shell model with spin–orbit coupling (see Portides 2011).

by construing their representational function as follows: a phenomenological model represents its target by providing a mechanism or an interaction of different mechanisms responsible for the observed behaviour of
the system, provided that the mechanism is explanatory of the observed
behaviour of the system.[6] By understanding scientific representation in this
manner we still rely on Goodman's suggestion that a phenomenological
model X is a Y-representation of its target Z, as long as Y is understood
to be a description of a mechanism or interaction of several mechanisms.
Hence, elements of the model denote putative entities and activities of the
target Z, and the mechanism picked out by the model is explanatory of
the observed behaviour of Z.

This understanding of scientific representation does not tie it just to
denotation but also to the explanatory function of models. According
to several authors (see, in particular, Morrison 1999 and Portides, 2011)
the primary function of scientific models is to explain the behaviour of,
and to operate as sources of knowledge about, target physical systems and
their constitutive parts. Morrison's argument, in which the representational capacity is correlated to the explanatory power of models, is meant
to point to the reason why a model is representational, namely because it
has explanatory power. The latter is a common feature of every representational model; hence, it is a necessary characteristic for a model to be representational. The notion of 'explanatory power' is admittedly too general,
but in the context in which I employ it we can avoid some ambiguity: the
more detailed the description of the postulated mechanism denoted by
the elements of the model the higher its explanatory power. The degree of
explanatory power of a model is determined by the level of specificity of
the mechanism it represents.

Let me elaborate with an example from nuclear physics, to help clarify the idea that phenomenological models represent by denoting elements of mechanisms that are explanatory of the target's behaviour. The
liquid-drop model of the structure of the nucleus, proposed in 1936 by
Niels Bohr (see Moszkowski 1957; von Buttlar 1968), is based on the analogy that the mean free path of nucleons must be significantly small compared with the nuclear radius, just as the mean free path of molecules in
a liquid drop is small compared with the radius of the drop. According
to the model, any energy acquired by a nucleon is quickly shared, giving

---

[6] The general characterization of mechanisms due to Machamer et al. (2000) seems suitable for my
purposes in this chapter: 'Mechanisms are entities and activities organised such that they are productive of regular changes from start or set-up to finish or termination conditions' (2000, 3).

rise to nuclear excitations that involve collective displacements of nucleons. Subsequently, the motions of individual nucleons are omitted and
the nuclear wave function is entirely described in terms of the position
of the nuclear surface. The energy equation is set up by employing three
primary idealizing classical assumptions: (1) that the nucleus in its stable state has spherical shape; (2) that for small deviations from sphericity,
where the surface undergoes deformation oscillations at constant density,
the surface tension of the nucleus acts as a restoring force; (3) that the
energy of the nucleus is the sum of the volume energy, surface energy, and
Coulomb energy and that, on the assumption of incompressibility, the
volume energy is independent of the nuclear shape, the surface energy is
least for spherical shape and increases with deviation from sphericity, and
the Coulomb energy decreases with deviation from spherical symmetry.
The result is a classical energy function for the collective motion:

$$H = E(0) + \sum_{\lambda} \sum_{\mu} \left(\tfrac{1}{2} B_\lambda \left| \dot{\alpha}_{\lambda\mu} \right|^2 + \tfrac{1}{2} C_\lambda \left| \alpha_{\lambda\mu} \right|^2 \right) \tag{3}$$

where $E(0)$ is the energy for spherically symmetric shape; C_λ are nuclear-
deformation-resistance coefficients; the quantities B_λ are mass parameters;
and $\alpha_{\lambda\mu}$ are deformation functions. Then the equation of motion is quan-
tized by introducing momenta $\pi_{\lambda\mu}$, canonically conjugate to the $\alpha_{\lambda\mu}$, so
that the quantum-mechanical Hamiltonian operator takes the form:

$$H = E(0) + \sum_{\lambda} \sum_{\mu} \left(\tfrac{1}{2} \frac{\left| \pi_{\lambda\mu} \right|^2}{B_\lambda} + \tfrac{1}{2} C_\lambda \left| \alpha_{\lambda\mu} \right|^2 \right) \tag{4}$$

Although the liquid-drop model is what we would call a semi-classical
model, it served an explanatory function primarily in our understanding of nuclear fission and the electric quadrupole moments of nuclei, and
eventually became an essential ingredient in the development of a more
elaborate nuclear model known as the unified model of nuclear structure.

To understand how the model postulates a mechanism that is explanatory of nuclear fission we must examine the primary guiding principle
that led to the construction of the model. In the early 1930s, before the
proposal of an adequate nuclear model, with the development of mass
spectroscopy it was found that the nuclear mass is related to the masses
of its constituent particles and to the nuclear binding energy B (i.e. the
minimum energy required to completely separate the nucleus' component

nucleons): $M_{nucl} = ZM_p + NM_n - c^{-2}Bn$. This result showed that nuclear binding energies are sufficiently large to affect nuclear mass. Another result with respect to nuclear binding energies was their approximate constancy for different nuclei (except the lightest). Along with other experimental results, these led in 1935 to von Weizsäcker's semi-empirical result regarding the binding energy of the nucleus. His semi-empirical mass formula consists of five components:

$$B = C_{vol} A - C_{surf} A^{2/3} - C_{coul} Z^2 A^{-1/3} - C_{sym} (A-2Z)^2 A^{-1} - C_{pair} A^{3/4} \delta \quad (5)$$

where Z is the proton number; and A is the total nucleon number. The rationale behind Weizsäcker's formula is this. The first three terms are just of the form suggested by the classical analogy with the charged liquid drop (assumption (3) above). If we consider an infinitely extendible liquid (of constant density) then the energy would be proportional to the number of particles. In the nuclear analogy, this volume energy is the average energy due to saturated bonds between the nucleons, which increases B. But since the nucleus is finite, the nucleons near the surface should interact with fewer nucleons (i.e. there should be unsaturated bonds). Thus B should decrease by an amount proportional to the surface area, i.e. to $A^{2/3}$. Furthermore, the binding energy reduces more on account of the Coulomb repulsion between any two protons. This is inversely proportional to the distance between two protons, which turns out to be inversely proportional to $A^{1/3}$. At this point the classical analogy ceases to help, and the following two considerations suggest the addition of the last two terms in Equation (5). The tendency of nuclei to have equal numbers of protons and neutrons N gives rise to the symmetry term which for $Z = N$ diminishes. Also a pairing term must be added to reproduce the special stability of even–even nuclei and the almost complete absence of stable odd–odd nuclei. Thus in the Weizsäcker formula, $\delta = +1$ for odd–odd nuclei, $\delta = 0$ for odd-nucleon nuclei, and $\delta = -1$ for even–even nuclei.

The liquid-drop model is a valuable tool for *explaining* the Weizsäcker semi-empirical formula, despite the fact that more detailed models are required to relate the magnitudes of the various terms to the basic interactions between nucleons. Since Weizsäcker's semi-empirical result is successful in yielding relatively accurate values for B and in reproducing all important nuclear trends, except for the lightest nuclei, the fact that some of its terms are explained by the liquid-drop model is regarded as an indicator of the success of the latter. This led to research in pursuing the consequences of the model. One notable result was the success of the model in providing the mechanism for explaining the phenomenon of nuclear fission of heavy elements.

Here is a sketch of the mechanism the model provides. We assume the nucleus to be roughly like a classical charged liquid drop. Nuclear matter is therefore assumed to be incompressible, just as a liquid almost is, but deformation is possible. If a spherical nucleus is deformed into an elongated shape its surface area would be increased. This reduces the surface energy compared with that of the spherical nucleus. In addition, the two halves of the elongated nucleus are now found at a larger distance apart, compared with the spherical shape. According to the model, in such a situation the following things would happen. First, the Coulomb repulsion is diminished because the average distance between protons increases. Second, the surface energy increases because the surface area increases. These two changes have opposing effects on the magnitude of the binding energy. Thus when the nucleus is disturbed and its shape changed, two obvious things can happen. The nucleus might return to its original shape, or it may undergo fission. The first happens when the charge density is small compared with the surface energy, and thus outweighs the Coulomb repulsion. In such cases, the nucleus tends to minimize its surface at all deformations, and the spherical shape is stable for every distortion up to a limit. Thus since, for light nuclei, the surface tension is more significant, the spherical shape is the stable configuration. But what happens in the limiting case where the spherical nucleus is so highly charged that the slightest displacement from its spherical shape will produce fission? The Coulomb repulsion dominates over the surface energy that tries to bring the shape back to spherical, and the two parts of the drop will be violently repelled from each other. This is so because the Coulomb energy increases with Z^2, whereas the surface energy increases with $A^{2/3}$; hence for a large Z the Coulomb energy prevails. The Coulomb repulsion is therefore responsible for making the nucleus move apart violently and release energy. The two changes, which have opposing effects on the magnitude of the binding energy, mean that heavy nuclei will demonstrate instability against deformation. A deformation of a large nucleus, whether spontaneous or initiated by the capture of a particle, may therefore lead to a large deformation and subsequently to a violent split-up into two or more parts of comparable mass and a release of energy. The liquid-drop model does not just offer this qualitative explanation for nuclear fission but it also provides, to a first approximation, good quantitative results. Being semi-classical, this construal of fission neglects two important quantum-mechanical effects: (1) fission may take place for excitation energies below the fission threshold by means of the tunnelling effect; (2) the vibration of the nucleus in the distorted mode will have zero-point energy.

Even though some quantum-mechanical effects are not explained and some important properties of nuclei are not adequately accounted for by the model (e.g. the special stability demonstrated by some nuclei with particular numbers of protons or neutrons, known as the 'magic-number' nuclei), the important point in discussing the liquid-drop model is that the model's representational success is measured by its ability to provide an explanatory mechanism for the phenomenon of nuclear fission. The Weizsäcker formula, which is important in the construction of the model, is something that must be explained by the model since the former acts as a mediator between model predictions and experimental data. The liquid drop provides an explanation – however incomplete – for three of the causal terms of the Weizsäcker formula by providing a description of the physical mechanism that represents the nucleus and gives rise to those terms of the formula, thus it represents some of the factors responsible for the effect of nuclear fission. Even though the liquid-drop model does not offer a comprehensive understanding of nuclear behaviour, its function is to provide an explanation of the phenomenon of nuclear fission. Hence, the primary purpose for discussing the model is not to argue about its infallible predictive and elaborate explanatory power; it is to show that the representational capacity of the model derives from its ability to provide a putative mechanism for the particular target's behaviour, for which it was built in the first place.

Seventy-five years after its inception one can argue that the liquid drop is unrealistic or far from the truth, hence we should exclude it from our attempts to draw general conclusions about scientific modelling and scientific representation. Such an argument could be based on either the semi-classical origins of the model or on the fact that since its inception models of nuclear structure that have a closer bond to quantum theory and yield more accurate predictions have been developed. Such an argument can be challenged in two ways. The first is that the argument misses the point about scientific representation. Scientific representation is distinguishable from 'getting it right'; it is about providing the right tools by which we can understand how physical systems work and predict their behaviour. This may mean that eventually we do get it right but this is a question about the truth of our scientific propositions and it should not be tangled up with our attempt to understand how our models represent. In this chapter I focus only on addressing what it means to provide the right tools by which to gain knowledge about the workings of physical systems, and the answer I offer to this is: to construct models whose elements denote entities and activities of putative mechanisms to explain a system's behaviour. It is an answer that only concerns what scientific representation is, not whether scientific inferences lead to approximately true propositions or not. The second challenge relies on the history of nuclear models. After almost eighty years of work in nuclear physics we can claim with confidence that what the liquid-drop model denotes, i.e. nucleons and strong interaction forces that hold the nucleons together, is not exactly right. Nevertheless, the liquid-drop model still has an important place in nuclear physics, although it has been significantly modified over the course of nuclear physics' history. In fact, the mechanism it provided served as the basis for the construction of more detailed mechanisms that more refined and comprehensive models of nuclear structure, such as the unified model, provide for the explanation of nuclear fission. These reasons make our attempts to understand how phenomenological models represent all the more important.

Could the view of scientific representation that I defend also do justice to representation by means of theory-driven models? If some theory-driven models can be used for the representation of physical systems then these particular models must have something in common with phenomenological models used for the same purpose. Of course, there is a striking difference between the two kinds of models that has been pointed out by a number of authors (see Cartwright 1999; Morrison 1999). The former are constructed by the use of the abstract theoretical calculus, whereas the latter are constructed by the use of more concrete concepts. Yet they both demonstrate some degree of representational capacity, albeit different. What characterizes the way theory-driven models are used for representational purposes is that they provide the source for building a model that captures those exact same things that phenomenological models are developed for, i.e. the particular physical constraints that affect the behaviour of respective target systems. In the case of the pendulum example, the LHO can function as the basis by which all known physical constraints can be absorbed into the model. The result of this absorption is what I have called the DO. When theory-driven models can function in this way, then they are representational candidates for specific targets. The LHO, for example, is a representational candidate for physical systems such as the pendulum, the mass-spring system, and the torsion pendulum. I think the important question to ask is this: What makes some theory-driven models capable of absorbing the specific physical constraints exerted on some physical systems in contrast to other theory-driven models that are incapable of demonstrating this characteristic?

I think the answer to this also lies with the notion of mechanism, in particular with what Machamer et al. (2000) call a mechanism schema. According to Machamer et al. (2000, 15) a mechanism schema is an abstract description of a type of mechanism, which is usually given in truncated form that can be filled in with descriptions of known component parts and activities. The LHO can be interpreted as describing exactly such an abstract mechanism schema that partially characterizes the behaviour of systems such as the pendulum, the torsion pendulum, and the mass spring. When the known constituent parts and activities are incorporated into the LHO, we are led to a specific instance of a mechanism or a specific instance of interactions of different mechanisms. Moreover, as more of the details of the known particular constituent parts and activities are incorporated into the model the latter's explanatory power increases. For example, the LHO explains only what Newton's laws dictate, namely that if the only forces acting on the pendulum are gravity and the tension due to the string then the bob will oscillate back and forth with constant amplitude of oscillation. Once the medium, inside which oscillations take place, is added and the activities it gives rise to are incorporated into the model then a number of other observable things are also explained. The change in the amplitude of oscillation is explained. The partial rotational motion of the bob is explained. And so on. One could give an entire list of observable events explained by the sixteen factors of the DO and their interactions. The point is that when we move from representing the general, i.e. a type of mechanism, to representing specific mechanisms or interactions of mechanisms then we gain in explanatory power. A theory-driven model represents a system as if aspects of its observed behaviour were due to a type of mechanism; it does that by denoting the general features of a type of mechanism. But a condition has to be fulfilled if such a model can be representational of a physical system: when the physical constraints of the actual system are considered and the descriptions of known component parts and activities are absorbed in the theory-driven model, then we are led to a model of greater explanatory power.

Phenomenological models, as well as theory-driven models that are supplemented with phenomenological elements (e.g. the DO model), pick out aspects of putative mechanisms for the observed behaviour of target systems. The difference between such models and theory-driven models lies in the fact that the former pick out particular mechanisms, whereas the latter pick out mechanism schemata. As descriptors of mechanism schemata, theory-driven models often are part of the heuristic that leads to a representation of a physical system. Occasionally they guide us towards the mechanism responsible for the observed behaviour without losing their status of being an aspect of the mechanism, as when we fill in the specific details of the pendulum that are left out of the LHO to be led towards the DO. Other times (especially in quantum-mechanical modelling) they are part of the heuristic that leads to a model that picks out a putative mechanism but they are discarded once the mechanism is in place.[7] Whichever the case, those theory-driven models employed in the construction of representational models are employed in the first place because they can be used to represent mechanism schemata that seem helpful in revealing the particular mechanisms responsible for the behaviour of physical systems.

5 CONCLUSION

My argument is that an important characteristic of scientific modelling is that some models represent the general aspects of their targets while others represent the more specific aspects as well. To accommodate this characteristic of models and maintain denotation at the core of representation we should construe scientific models as being representations of putative mechanisms or putative mechanism schemata. This view of scientific representation construes a model X as a mathematical entity whose terms denote component parts and activities of target systems. Furthermore model X integrates these elements in ways that provide explanations for the observed behaviours of its target. The degree of explanatory power of a model is determined by the level of specificity of the mechanism it represents. In a nutshell, a scientific model represents its target system as if its observed behaviour is produced by an underlying mechanism or interactions of several mechanisms, so long as these mechanisms are explanatory of the target's behaviour.

This is an account of scientific representation that distinguishes it from other kinds of representation. We use symbols to represent targets in all sorts of domains. Sometimes the representational vehicle we use is highly complex. Such is the case with musical symbols. Various symbols and their configuration on a staff are used to represent music notes, rests, breaths,

---

[7] For a detailed account of such a case see Portides 2011.

caesuras, pitch, pitch range, accidentals, time signatures, dynamics, articulations, octaves, and so forth, and various subdivisions of the above. They are also used to represent how the various bits of this notation are integrated together into one whole. We create a highly complex representational system that exhibits the characteristic of denotation. Nonetheless, such systems do not need to denote aspects of mechanisms and more importantly their function is not explanatory. Furthermore, such representations do not divide into general and specific categories, because their denotation is fixed, i.e. it is not subject to varying idealizing assumptions. However, if scientific models are denotative representational vehicles and we want to capture the general/specific divide, then their representational and explanatory functions should be linked. This is what makes scientific models distinct forms of representation.

REFERENCES

Cartwright, N. D. (1999) Models and the limits of theory: Quantum Hamiltonians and the BCS models of supercondictivity. In Morgan and Morrison 1999, pp. 241–281.
da Costa, N. C. A. and French, S. (2003) *Science and Partial Truth*. Oxford University Press.
Elgin, C. Z. (2009) Exemplification, idealization, and scientific understanding. In M. Suárez (ed.) *Fictions in Science*. New York: Routledge, pp. 77–90.
Goodman, N. (1976) *Languages of Art*. Indianapolis: Hackett Publishing Co.
Hartmann, S. (1999) Models and stories in hadron physics. In Morgan and Morrison 1999, pp. 326–346.
Hughes, R. I. G. (1997) Models and representation. *Philosophy of Science* 64 (Proceedings): S325–S336.
Machamer, P., Darden, L., and Craver, C. F. (2000) Thinking about mechanisms. *Philosophy of Science* 67: 1–25.
McMullin, Ernan (1985) Galilean idealisation. *Studies in History and Philosophy of Science* 16: 247–273.
Morgan, M. S. and Morrison, M. (eds.) (1999) *Models as Mediators: Perspectives on Natural and Social Science*. Cambridge University Press.
Morrison, M. (1997) Models, pragmatics and heuristics. *Dialektik* 1: 13–26.
     (1999) Models as autonomous agents. In Morgan and Morrison 1999, pp. 38–65.
Moszkowski, S. A. (1957) Models of nuclear structure. In S. Flügge (ed.) *Encyclopedia of Physics*, vol. XXXIX: *Structure of Atomic Nuclei*. Berlin: Springer, pp. 411–550.
Nelson, R. A. and Olsson, M. G. (1986) The pendulum – Rich physics from a simple system. *American Journal of Physics* 54, no. 2: 112–121.

Portides, D. (2011) Seeking representations of phenomena: Phenomenological models. *Studies in History and Philosophy of Science*, Part A, 42, no. 2: 334–341.

Suárez, M. (2004) An inferential conception of scientific representation. *Philosophy of Science* 71: 767–779.

van Frassen, B. C. (1980) *The Scientific Image*. Oxford: Clarendon Press.

von Buttlar, H. (1968) *Nuclear Physics*. New York: Academic Press.

CHAPTER 12

Referring to localized cognitive operations in parts of dynamically active brains

William Bechtel

The project of referring to localized cognitive operations in the brain has a long history and many impressive successes. It is a core element in the practice of giving mechanistic explanations of mental abilities. But it has also been challenged by prominent critics. One of the critics' claims is that brain regions are not specialized for specific cognitive operations and any science that refers to them is misguided. Most recently this claim has been advanced by theorists promoting a dynamical-systems perspective on cognition. There are, however, two ways to view the dynamical-systems perspective. The first is as a competitor to the mechanist perspective, rejecting altogether the conception of the brain as a mechanism or set of mechanisms underlying mental phenomena and thereby rejecting any reference to localized cognitive operations. The second is as a corrective to an overly simplistic conception of a mechanism and as complementary to a more adequate understanding of how mechanisms function. In this chapter I defend the latter point of view. On this point of view, the traditional project of referring to localized mental operations in the brain is still important, but both the cognitive operations and brain regions in which they are localized must be conceived in the context of a dynamically active system.

In the first section, I describe the traditional project that refers to localized cognitive operations in the brain, and situate it within the framework of mechanistic explanations of psychological phenomena. On this view, brain regions are construed as parts of a mechanism that are specialized for specific (information) processing tasks and perform these tasks when called upon by appropriate initiating conditions. I characterize the accompanying conception of brain regions in terms of Simon's conception of hierarchically organized, nearly decomposable components. I also briefly describe the opposition of some contemporary dynamicists who reject the very project of parceling of the brain into regions and referring to them as performing distinct operations.

In the second section I develop a different perspective on brain regions, one that construes them as active components in a dynamically self-organizing system. On this account, the individual regions of the brain are endogenously active and as a result of this activity organize into specialized processing components. Yet even as the regions of the brain specialize, they remain integrated with other regions in a mode of organization known as a *small-world network*. In such organization local clustering gives rise to specialized regions but long-range connections link processing within these regions to activity elsewhere in the system, allowing activity elsewhere to modulate the behavior of local clusters. I present several sources of evidence that the brain exhibits small-world properties at multiple levels of organization.

In the final section I return to the question of referring to localized cognitive operations in brain regions and first examine evidence that small-world organization is also exhibited in the endogenous functioning of the brain and that functionally characterized small-world networks correlate with those characterized structurally. As a result, neither the decomposition of the mechanism into parts nor operations yields the sort of distinct parts or distinct operations commonly assumed in a mechanistic account, but rather parts and operations that are highly integrated with each other in ways that undermine traditional approaches to linking well-specified operations to clearly differentiated parts. Instead, the operations performed by a part must be recognized to vary substantially depending on their interactions with other parts. To understand this conception of the brain, a different conception of mechanistic explanation is required, one I refer to as *dynamic mechanistic explanation*. Dynamic mechanistic explanations are still mechanistic, and so make reference to operations localized within parts, but respect the dynamic processes that require characterizing both parts and operations relationally in terms of how they are situated in endogenously active dynamic networks.

I   THE TRADITIONAL APPROACH TO AND CRITICISM
OF REFERRING TO LOCALIZATION OF MENTAL
OPERATIONS IN BRAIN MECHANISMS

Although claims that particular brain regions are responsible for specific mental phenomena were advanced early in the development of neuroscience (most notoriously by Gall around 1800), researchers only began to utilize naturally occurring or experimentally induced lesions or electrical stimulation to link mental processes experimentally to specific brain

regions in the second half of the nineteenth century. Especially noteworthy accomplishments were Broca's (1861) localization of articulate
speech in the region in left prefrontal cortex that came to bear his name
(based on deficits exhibited in patients with lesions to this region), Fritsch
and Hitzig's (1870) localization of motor control of muscle contractions to
regions of the motor strip (based on responses after mild electrical stimulation to the cortex in dogs), and Munk's (1881) localization of basic visual
processes to the occipital lobe (based on deficits following lesions to the
occipital cortex in various species).

    These successes seemed to answer the skeptical worries about localiza-
tion advanced by Flourens (1824) in criticizing Gall. In his lesion experiments on pigeons, Flourens had succeeded in differentiating functions
between the brain stem, cerebellum, and cerebrum, but failed to find specialized regions for perception, memory, or problem solving within the
cerebrum. Rather, he claimed that lesions to the cerebrum generated deficits in proportion to how much was removed, and he embraced a holist
perspective of the cerebrum. In the wake of the success of Broca, Fritsch
and Hitzig, and Munk, it was no longer plausible to reject all localization
of function in the cortex, but many investigators adopted the perspective
advanced by Lashley (1950) – primary sensory and motor processing was
localized, but further processing and especially formation of memories
depended on generalized *association* areas which operated by a principle of
mass action, whereby performance decreased proportional to the amount
of cortex lost, with no specific deficits corresponding to particular losses.
On Lashley's view, it was a mistake to refer to specific cognitive functions
performed in these brain areas. Although lesion studies contributed to
answering Lashley, further successes in localizing information processing
in the brain relied heavily on the development of techniques for recording
action potentials from individual neurons and correlating the increased
generation of action potentials in particular neurons with the presentation
of specific stimuli. After Hubel and Wiesel (1962, 1968) showed that the
cells in the occipital cortex that Munk had viewed as the locus of visual
processing seemed only to differentiate edges, they and other researchers
looked more anteriorly and identified areas in temporal and parietal cortex responsive to other features of visual stimuli, including motion, color,
shape, and object identity.

    I have analyzed the history of localizing different steps in processing
visual inputs in different cortical regions elsewhere (Bechtel 2008). What
is important for present purposes is the conception of brain processing assumed in this research. The assumption is that the brain contains

a mechanism for processing visual inputs and that the goal was to discover this mechanism. A mechanism for these purposes is a system that consists of distinguishable parts that perform specific operations which are organized to generate the phenomenon of interest (Bechtel and Richardson [1993] 2010; Bechtel and Abrahamsen 2005; Machamer et al. 2000). Differentiating parts and operations required the development of appropriate research techniques to decompose the brain structurally and functionally. Using staining techniques to differentiate the distribution of neurons in the cortex, for example, Brodmann ([1909] 1994) and his contemporaries developed maps of different brain regions early in the twentieth century. These maps were further differentiated using such tools for identifying patterns of connectivity between neurons and functional maps (Felleman and van Essen 1991). It was by recording from these areas as different stimuli were presented to an animal (typically a macaque monkey) that researchers localized various stages in the processing of visual information in the cortex.

Implicitly, in differentiating brain areas and referring to functions localized in them, neuroscientists adopt the conception of a *hierarchically organized nearly decomposable* system articulated by Simon (1962). A fully decomposable system (which Wimsatt 1986 characterizes as an aggregative system) is one in which each component functions independently of the others, and the whole is just the collection of separate components. Because the components do not cooperate to accomplish anything, such a system is not a mechanism. In a nearly decomposable system, the components perform independent functions but interact in ways that are 'not negligible'. Simon characterizes such nearly decomposable systems in two propositions: '(*a*) in a nearly decomposable system, the short-run behavior of each of the component subsystems is approximately independent of the short-run behavior of the other components; (*b*) in the long run, the behavior of any one of the components depends in only an aggregate way on the behavior of the other components' (Simon 1962, 474). Simon illustrates such organization by describing a building that is thermally insulated from the environment and is divided into rooms that are well, but not perfectly, insulated and that initially vary in temperature. In the short run, each room goes to thermal equilibrium, little affected by the other rooms. More gradually the whole building reaches thermal equilibrium as heat dissipates from warmer to cooler rooms. The equations describing the initial changes need only employ variables for properties of the local room (as those of the other rooms will have negligible effects on it) and the equations for the second round of changes need only include variables for mean values of the rooms. Applied to the brain, in the short term the processing within local brain regions will proceed, affected little by processing in other regions, whereas over the longer term the completed processing in one area will affect that in other areas to which it is connected.

Simon offers two sorts of arguments for the ubiquity of nearly decomposable systems, especially in living organisms. First, they are the most likely to evolve. He refers to the subsystems within nearly decomposable systems as *stable subassemblies* and illustrates the advantage of building larger systems from stable subassemblies with a tale of two watchmakers, Tempus and Hora. Each makes equally fine watches consisting of approximately 1,000 parts that take nearly a day to assemble. Without stable subassemblies, Tempus has to position all 1,000 pieces before his watches are stable. Hora builds his watches by first building assemblies of about 10 pieces that remain stable, and then combining about 10 of these into further stable assemblies, and finally combining these into the whole watch. With even a modest rate of interruptions (1 every 100 steps of assembly), Tempus will only produce a stable product 44 times per million attempts, while Hora will complete an assembly before interruption 90 percent of the time. Simon extrapolates this lesson to biological evolution, arguing that if more complex biological systems evolved from stable simpler systems, natural selection would be able to generate a complex life form far faster than supposed in various anti-evolutionist criticisms. Simon also argues for the hierarchical organization of nearly decomposable systems by suggesting that only such systems would be intelligible to us:

> If there are important systems in the world that are complex without being hierarchic, they may to a considerable extent escape our observation and our understanding. Analysis of their behavior would involve such detailed knowledge and calculation of the interactions of their elementary parts that it would be beyond our capacities of memory or computation. (1962, 477)

Although not strictly required by the basic conception of a mechanism or by Simon's conception of hierarchically organized nearly decomposable systems, a natural assumption is that a mechanism built in such a manner will function through a sequence of specialized operations, each of which makes its product available for further processing by components conceived as downstream of it. Thus, mammalian visual processing is assumed to begin with processing by the retina and continue through the LGN (lateral geniculate nucleus) to V1 (primary visual cortex), V2, etc. This conception was slightly complicated by the discovery of two different pathways from V1, one culminating in the medial temporal cortex and involved in object recognition and a second culminating in the parietal cortex and involved either in identifying where the stimulus is located (Ungerleider and Mishkin 1982) or the motor response one might make to it (Milner and Goodale 1995). As there are projections between the pathways at various stages of processing, van Essen and Gallant (1994) propose we think of them as streams which, while mostly flowing in isolation, sometimes send tributaries to the other. They, however, retain the sequential perspective, which is indeed highly natural for humans trying to understand the operation of a mechanism. We think of a mechanism as doing something (e.g. locating an object in the space around us) and we investigate how it performs the sequence of operations needed to do so. Machamer et al. (2000) in fact enshrine the perspective in their definition of a mechanism as 'productive of regular changes from start or set-up to finish or termination conditions' (3).

As I noted at the outset, this classical conception of the brain as a mechanism has been challenged by theorists who advocate a dynamical-systems approach to brain function. This approach characterizes brain function in terms of reciprocal causal interactions whereby 'each and every component of a system contributes to every behavior of the whole system' (van Orden and Paap 1997, S92). While it might seem that such a holistic perspective would provide no tools for understanding specific activities, dynamical-systems theory provides a variety of ways to identify different patterns of relations between variables characterizing the system (e.g. by describing trajectories in the state-space defined in terms of these variables and the patterns of these trajectories) that can be related to different cognitive activities. The resulting approach emphasizes patterns in the global behavior of the brain that correspond to particular cognitive activities while eschewing any reference to these cognitive activities being localized in regions of the brain. In arguing for this approach as preferable to the mechanistic one of localizing operations within the brain through techniques such as neuroimaging (van Orden and Paap 1997; Uttal 2001) or analysis of lesions (van Orden et al. 2001), dynamicist critics emphasize the persistent variability in the data appealed to in localization studies; while this variability is usually treated as noise, the critics see it as an indicator of the flawed conception of the brain that underlies localization.

In the following sections I propose to reconcile the seemingly incompatible localizationist, mechanistic perspective and the dynamical-systems approach, a reconciliation that requires reconceptualizing the project of localization. This reconciliation will retain the appropriateness of reference to localizing functions in the brain but significantly revise both what a brain region is taken to be and what is taken to be localized in one. I begin by focusing on how recent research on dynamical activity in the brain requires reconceiving brain regions.

II   RECONCEIVING BRAIN REGIONS FROM A DYNAMICAL, INTERACTIVE PERSPECTIVE

As plausible as Simon's account of hierarchically organized nearly decomposable systems and Machamer et al.'s conception of the sequential operation are as ways of understanding mechanisms, they fit poorly with the modes of organization increasingly being identified in actual biological systems. Within the visual system, and indeed in the brain generally, there are as many projections backwards through the presumed processing stages as there are forward projections, and there are even more lateral projections within and between brain regions (Lorente de Nó 1938). While widely acknowledged, these projections are often neglected in functional accounts since it is not clear how they promote the presumed processing ends of the system. One of their effects, as we will see, is to undercut the view of brain regions as nearly decomposable parts while still allowing for localized regions that perform specialized operations.

To motivate reconceptualizing brain regions, consider again Simon's proposal that complex systems are constructed from stable subassemblies. Such assembly of independent components into larger systems appears to occur infrequently in evolution and even when it does, the result is an integrated, and so less decomposable, system. One of the best-supported proposals for such an evolutionary process hypothesizes that mitochondria in animals and chloroplasts in plants arose when one prokaryotic organism engulfed another to form eukaryotic cells. Mitochondria on this account resulted from the incorporation of proteobacteria (probably purple non-sulfur bacteria) while chloroplasts resulted from the incorporation of cyanobacteria.[1] In the case of mitochondria, the host that had previously relied on glycolysis for energy acquired from the symbiont the

---

[1] This proposal, known as the endosymbiotic theory and commonly associated with Lynn Margulis (Sagan 1967), was initially suggested (in the case of chloroplasts in plants) by the German botanist Andreas Schimper in 1883, further developed by the Russian botanist Konstantin Mereschkowsky in 1905, and extended to mitochondria by Ivan Wallin in the 1920s. It was, however, largely ignored until electron microscopy provided further evidence for the similarities between chloroplasts and cyanobacteria and Stocking and Gifford demonstrated the occurrence of DNA in mitochondria. Margulis (1981) extended the endosymbiotic theory to explain the origins of flagella and cilia, but these proposals are harder to support since flagella and cilia do not have their own DNA.

ability to perform oxidative phosphorylation, thereby extracting significantly more energy (in the form of ATP) from its food sources. Although
the symbiont retained its DNA and ribosomes, and replicates through a
process of division (which provides important evidence for its origin as
an independent organism), it has nonetheless been highly integrated into
the life of the host cell (e.g. some of its DNA has been transferred to
the host's nucleus, and the symbiont has become dependent on proteins
generated by the host and the conditions within it). Thus, even when separate components, originally capable of functioning independently, are
brought together, over time they integrate their operation and become far
less decomposable.

More commonly, however, evolution proceeds by expanding, often by
duplicating (e.g. through an extra round of cell division during development) existing parts of the mechanism and then allowing structural and
function differentiation within the system. That is, component parts and
operations arise from specialization of regions within an initially more
homogenous part of the system. I will offer a sketch of such a specialization process in the context of introducing a mode of organization that has
only been seriously explored in the past two decades but is increasingly
recognized as extremely common in evolving systems, including living
organisms – small-world networks. Small-world networks contain specialized components, but they are not nearly as independent in their operation as the subsystems in a hierarchically organized nearly decomposable
system.

Through much of the second half of the twentieth century, mathema-
ticians in the subfield of graph theory focused on three designs for networks: regular lattice, totally connected, and randomly connected. Two
measures particularly useful in understanding network functioning are
its clustering coefficient and its characteristic path length (Bullmore and
Sporns 2009). The clustering coefficient (C_j) for a given node j measures
the percentage of the possible connections between neighbors of j that are
actually realized. A high clustering coefficient for a whole network (C)
indicates that components are highly connected to those in their neigh-
borhood, a necessary condition for them to cooperate in performing specialized processing. The characteristic path length (L) is the mean of the
shortest path lengths between nodes, where path length is measured in
the number of connections that must be traversed. A low L enables high
integration of the activity of all units in the network. In a regular lattice,
nearby units are connected, resulting in a high value for C, but a large
number of connections must typically be traversed to go from a given

unit to a specific other unit, resulting in a high value for L (Simon's nearly decomposable systems are similar in this respect to regular lattices). In a network in which connections between units are determined randomly, if there is a path between two units, there will likely be a relatively short path between them (hence a low value for L). However, there will also be little clustering (low C) as the connections are as likely to be between distant as to be between nearby units.

Ideal conditions for information processing arise when C is high, enabling local clusters of units to work together, and L is low, enabling coordination between the clusters. This condition obtains in a fully connected network, but because all units are connected together there is little opportunity for specialization. It also obtains, though, in a form of organization known as *small world*, a concept introduced by Milgram (1967) on the basis of an experiment: Individuals in the United States were requested to forward a letter they received to someone they knew personally with the goal of eventually getting it to a designated individual in Cambridge, MA. Surprisingly, on average those letters that made it to the destination did so in less then six steps. This gave rise to the popular notion that on average six degrees of separation exist between any two human beings. Watts and Strogatz (1998) provided a conceptual grounding for this phenomenon when they characterized a class of networks that result from randomly reconnecting a few of the connections in a regular lattice, providing rapid long-distance connection across a large network. After rewiring only a relative small percentage of the connections, they obtained networks that still exhibited high C but greatly reduced values for L.

In addition to characterizing such a network structure, Watts and Strogatz also demonstrated that a variety of real-world networks, including networks of movie actors linked by coappearances, the electrical power grid of the western United States, and the neural network of the nematode worm *Caenorhabditis elegans*, exhibited the characteristics of a small world. Watts and Strogatz also examined the functional properties of small-world networks, showing how they allow for rapid spread of infectious diseases, enable efficient problem solving in cellular automata, and reduce the likelihood of cooperation in iterated prisoner's dilemma games. Of particular interest, they examined coupled phase oscillators and demonstrated that synchronization occurred almost as fast in small-world networks as in fully connected ones. They commented: 'This result may be relevant to the observed synchronization of widely separated neurons in the visual cortex if, as seems plausible, the brain has a small-world architecture' (1998, 442).

Subsequent research has indeed demonstrated a small-world architecture in the mammalian cortex, including the visual system. Sporns and Zwi (2004) developed connectivity matrices based on data from published studies of neuroanatomy – Felleman and van Essen's (1991) study of the macaque's visual cortex, Young's (1993) study of the whole macaque cortex, and Scannell et al.'s (1999) study of the cat cortex. Sporns and Zwi showed that these networks exhibit the small-world properties of high C and low L.[2]

Watts and Strogatz introduced a procedure for creating small-world networks by rewiring regular lattice networks to replace a small percentage of local connections with more distant ones. Such a process is similar to the proposal above that even if a complex system developed from combining previously independent stable mechanisms, over time the components would develop increased connections with each other. But Rubinov et al. (2009) offer an intriguing alternative account of how small-world organization might arise, one that fits with the process of specialization after replication described above. They start with the idea that neurons are not passive elements awaiting an input but endogenous oscillators. There is considerable evidence of spontaneous oscillations in ion concentrations across cell processes (as well as selective synchronized oscillations in response to oscillation of other neurons at a particular frequency), which can give rise to spontaneous spiking (Llinás 1988).[3] To analyze how such

---

[2] Other research has identified features characteristic of small-world organization, including reciprocal and clustered connections in local circuits in mammalian neocortex through multielectrode recording (Song et al. 2005). Very recently tensor diffusion imaging has enabled MRI studies to identify fiber bundles in the human brain, and they too provide evidence of small-world organization (Hagmann et al. 2008).

[3] The inquiries of Hodgkin, Huxley, and Katz in the 1940s and 1950s into the generation of action potentials revealed voltage-dependent membrane conductances in potassium and sodium. Investigations conducted in invertebrates, especially mollusks (Kandel 1976), revealed numerous other voltage-dependent conductances beyond those responsible for action potentials which eventually were found to figure in mammalian neurons. These discoveries, together with a developing appreciation of intracellular signaling processes and of the variety of neurotransmitters and neuromodulators operating across neurons, resulted in a much richer understanding of physiological processes in individual neurons. Among the critical discoveries was Llinás and Yarom's (1981) finding of low-threshold Ca^{++} conductances in neurons in the inferior olive in mammals, which is more active when the neuron is at its resting potential than when it is hyperpolarized. They showed that this conductance renders the neuron into a spontaneous oscillator or resonator (a cell that oscillates in response to endogenous oscillators at specific frequencies). The spontaneous spiking of inferior olive neurons generates synchronized activity in a wide population of Purkinje cells in cerebellar cortex, which then provide inhibitory synchronized inputs to cerebellar nuclear neurons. Their rebound responses activate motor neurons in a synchronized manner, generating the physiological tremor with a frequency of approximately 10 Hz. Llinás (1988) draws out broad consequences of the spontaneous activity in the inferior olive: 'The organization of the IO nucleus demonstrates that the oscillatory properties of single neurons, arising from a congruous set of electrical events, can

behavior could generate small-world organization, Rubinov et al. employ a neural network developed by Gong and van Leeuwen (2004) that uses a logistic map as the basis for the activation function so that the activation value of a unit at the next time step $f(x)$ depends upon its current activation (x):

$$f(x) = 1 - ax^2$$

Under appropriate values of a (in their simulations they used 1.7), a unit can exhibit either regular oscillations or chaotic behavior. Gong and van Leeuwen then couple oscillators so that the activation of one depends upon, not just its previous activation, but those of each of the other units (j) to which it is connected:

$$x_{n+1}^i = (1 - \varepsilon) f(x_n^i) + \frac{\varepsilon}{M_i} \sum_j {}_{j \in B(i)} f(x_n^j)$$

Here ε determines how much a given unit is responsive to the previous activation of other units (in their simulations $\varepsilon = 0.5$). Such a network will generate complex dynamic activity in which various units will spontaneously synchronize their oscillations and subsequently spontaneously desynchronize. Gong and van Leeuwen then rewire the network by, at each time step, pruning the connection between a unit and another with which it is least synchronized, and adding a new connection to that unit with which it is not directly connected but with which it exhibits the greatest synchrony. From an initially random pattern of connectivity, such a network develops a small-world organization.

What is particularly intriguing about this model for the development of small-world organization is that it shows how structural organization can arise spontaneously, relying on endogenous activity within a network. In the model, the network was not employed in any information-processing task. Were the network employed in a task, though, the same processes of forming clusters integrated with other clusters in a small-world organization would be employed. The clusters formed could then subserve the relevant information-processing tasks. I turn to functional considerations in the next section.

The emerging conception of the structure of the brain that I have explored in this section departs significantly from Simon's conception of

activate a large number of neurons over a wide area. The ability to project oscillatory rhythms and to generate synchronous firing in large populations of cells may be one of the important properties of intrinsic electroresponsiveness' (1661). Llinás also reviews a vast array of other ionic conductances found in many neurons in the brain.

a hierarchically organized decomposable system and from the sequential conception of a mechanism presented above. Rather than a brain region that constitutes a component in a mechanism being largely isolated from others except for receiving the products of processing by other components and directing its products to other components, a brain region will be interconnected with numerous others. As a result, its behavior at any given time will depend, not just on its inputs and its internal constitution, but on the signals it receives from these other units. Such a perspective is suggestive of the holism that inspired skeptics of localization, but there is an important difference. The significance of clustering in small-world networks is that different parts of the network can still specialize for processing different kinds of information. Thus, referring to brain regions and localizing functions in them still makes sense, but the brain needs to be understood as involving components that are not operating in isolation but are highly sensitive to ongoing activity elsewhere in the mechanism.

III   RECONCEIVING LOCALIZATION FROM A DYNAMIC
MECHANISTIC PERSPECTIVE

In introducing their review of evidence for a small-world architecture in the cortex, Sporns and Zwi (2004) identify the dual role cortical connectivity plays in neural processing: 'First, it is critical in generating functional specificity (i.e. information) of local cell populations and areas within cortex. Second, it allows the integration of different sources of information into coherent behavioral and cognitive states' (146). I will briefly explore how both of these are manifest in brain activity, before turning to recent research using new approaches to fMRI (functional magnetic resonance imaging) that suggests how large-scale networks in the brain coordinate their functional activity through synchronization. I will then focus on the implications of these perspectives for the project of understanding the type of mechanism involved in cognition and how it supports referring to operations localized within such a mechanism.

The first functional consequence of small-world organization, functional specificity, results from the clustering of units into local subsystems. Even though many other components modulate the behavior of units in a local cluster, it is still possible, in part by focusing on the task the overall mechanism is performing, to identify and refer to the specific processing occurring in the local cluster. In particular, the traditional strategies for localization remain informative in characterizing the processing in local areas. The middle temporal visual area (MT) is appropriately characterized as playing a role in motion detection, as revealed in the range of techniques (lesion, recording, stimulation) that indicate its importance to that task (Britten et al. 1992). These strategies of referring to localized operations, however, should not be viewed as completing the inquiry, but only as a probative beginning. Once MT is identified as contributing to motion perception, researchers need to understand how its contribution is modulated by activity elsewhere in the brain.

The second functional consequence of small-world organization, integration into coherent global states, is illustrated at a coarse grain by the role oscillations in thalamic neurons play in producing global states such as attentive awakeness, drowsiness, and sleep, which modulate processing in many local circuits, including MT. When thalamic neurons are slightly depolarized, they oscillate at 10 hertz (Hz), but when hyperpolarized they oscillate at 6 Hz. Given the loops between thalamic and cortical neurons, the thalamus is able to differentially modulate cortical activity, depending on which of the two states it is in. Firing at 10 Hz, thalamic neurons act as relay elements communicating information to and from cortical neurons. Firing at 6 Hz, on the other hand, they serve to entrain cortical neurons, generating alpha rhythms or the pattern of spindling exhibited in early stages of sleep (Steriade and Llinás 1988). Motion detection in MT, as well as many other cortical operations, is modulated by the frequency of such oscillations.

Integration is achieved, not just at such a global level, but in more local circuits within the brain. One way these circuits have been revealed is through a relatively recent way of interpreting fMRI results. The BOLD (blood oxygen level dependent) signal employed in fMRI research registers the oxygen concentrations in the brain within areas that can be as small as 2 millimeters. In the most familiar applications of fMRI, changes in the BOLD signal are correlated with stimuli presented to or tasks performed by a person in the scanner (Posner and Raichle 1994). Well before this approach to fMRI studies was developed, however, researchers measured blood flow and brain activity when no tasks were being performed. Using the Xenon-133 inhalation technique to measure regional cerebral blood flow, Ingvar (1975) showed high levels of frontal activity when subjects were at rest and proposed that this activity reflected 'undirected, spontaneous, conscious mentation, the "brain work," which we carry out when left alone undisturbed'. Studies on resting brains were temporarily superseded once it became possible to measure blood flow during performance of cognitive tasks. They were revived, though, as Raichle et al. tried to understand why a range of brain areas seemed to be regularly deactivated in task situations – including the precuneus and posterior cingulate cortex, inferior parietal cortex, left dorsal lateral prefrontal cortex, a medial frontal strip that continued through the inferior anterior cingulate cortex, left inferior frontal cortex, left inferior frontal gyrus, and the amygdala. Raichle et al. proposed that these areas that are more active in the absence of task requirements and deactivated in task conditions constitute a *default network* (2001). Raichle and Mintun (2006, 463) provided further clarification of what they intend by a default network, 'We concluded these regional decreases, observed commonly during task performance, represented the presence of functionality that was ongoing in the resting state and attenuated only when resources were temporarily reallocated during goal-directed behaviors, hence our original designation of them as default functions.'

Characterizing these areas as a network requires more than establishing that they all exhibit less blood flow in task conditions – one must establish that the reduced activations are somehow coordinated. The avenue to doing this involved applying an innovative fMRI analysis pioneered by Biswal et al. (1995), who showed how to use fMRI time-series data to examine the microtemporal dynamics of blood flow. These investigators obtained BOLD signal values every 250 milliseconds after a subject performed a hand movement and identified low-frequency (less than 0.1 Hz) synchronized fluctuations in the left and right motor areas. Synchronizing oscillations requires the communication of a signal between the independent oscillators, indicative of a functionally integrated network of brain areas. Cordes et al. (2000) used a similar approach to identifying networks exhibiting synchronized blood flow within regions activated in sensorimotor, visual, auditory, and expressive and receptive language tasks. The technique of identifying networks by finding areas in which fluctuations in the BOLD signal are correlated is known as functional connectivity MRI (fcMRI).

Using fcMRI, Greicius et al. (2003) demonstrated, in the context of a working-memory task, 'a cohesive, tonically active, default mode network' (256) (cohesiveness refers to synchronized activity). It is noteworthy that synchronized oscillations in the default network are maintained even in task conditions, which results in lower overall activity in this network that is nonetheless still synchronized. One can then compare how oscillations in the regions constituting the default network relate to those in other brain areas. Fox et al. (2005) conducted an fcMRI analysis of imaging results during an attention-demanding task and found synchronized oscillations in a set of areas distinct from the default network, including the intraparietal sulcus, frontal eye field, middle temporal region, supplementary motor areas, and the insula. Oscillations in these areas were synchronized with each other, but not with areas in the default network. Fox et al. referred to this as anticorrelation and commented:

> anticorrelations may be as important as correlations in brain organization. Little has been said previously in the neuronal synchrony literature regarding the role of anticorrelations. While correlations may serve an integrative role in combining neuronal activity subserving similar goals or representations, anticorrelations may serve a differentiating role segregating neuronal processes subserving opposite goals or competing representations. (2005, 9677)

Subsequently, researchers have distinguished six anticorrelated networks (Mantini et al. 2007).[4]

Synchronization of processing in different brain regions has also been demonstrated in electrophysiological investigations measuring local field potentials in multicell recording or in EEG studies. The best known of these is the finding of synchronized oscillations in the range of 40 Hz in brain areas involved in representing different features of visual stimuli, which has been proposed to explain how the brain binds together different features of a visual stimulus as they are processed in different regions of the visual system (Gray and Singer 1989). The synchronized oscillations found with fcMRI are at a much lower frequency (<0.1 Hz) than those typically reported with local field potentials or EEG (usually in the 1–80 Hz range). However, oscillations at different frequencies may be related. Laufs et al. (2003) identified significant correlations between alpha (8–12 Hz) and beta oscillations (12–30 Hz) in EEG and BOLD fluctuations in specific brain networks. Mantini et al. (2007) related the six anticorrelated networks they found in fMRI with power fluctuations in all EEG bands, and concluded: 'Each brain network was associated with a specific combination of EEG rhythms, a neurophysiological signature that constitutes a baseline for evaluating changes in oscillatory signals during active behavior' (13170). For example, the default network showed positive correlations with alpha- and beta-band oscillations while the attention network exhibited negative correlations.

Going beyond mere correlations, a number of researchers have explored the potential causal relations between oscillations at different frequencies.

---

[4] Achard et al. (2006) applied wavelet analysis to time-series data derived from fMRI to estimate frequency-dependent correlation matrices for the whole human brain, and identified a small-world topology of sparse connections. They identified a small number of highly connected hubs located in multimodal association regions.

One relevant finding is that in the cortex of mammals, the amplitude (power density) of EEG oscillations is inversely proportional to their frequency ($1/f$). In addition, the phase of lower-frequency oscillations seems to modulate the amplitude of those at higher frequencies, which results in a nesting relation between the frequency bands. Lakatos et al. (2005) referred to this as the 'oscillatory hierarchy hypothesis' (see also Canolty et al. 2006). Since oscillations at lower frequencies tend to synchronize over more widely distributed brain regions than those at higher frequencies (Buzsáki and Draguhn 2004), this suggests that the relatively slow oscillations detected with fMRI might serve to modulate specific processing, reflected in the higher-frequency oscillations, in local brain areas:

The power density of EEG or local field potential is inversely proportional to frequency (f) in the mammalian cortex. This $1/f$ power relationship implies that perturbations occurring at slow frequencies can cause a cascade of energy dissipation at higher frequencies and that widespread slow oscillations modulate faster local events. These properties of neuronal oscillators are the result of the physical architecture of neuronal networks and the limited speed of neuronal communication due to axon conduction and synaptic delays. Because most neuronal connections are local, the period of oscillation is constrained by the size of the neuronal pool engaged in a given cycle. Higher frequency oscillations are confined to a small neuronal space, whereas very large networks are recruited during slow oscillations. (Buzsáki and Draguhn 2004, 1926)

In this section so far I have focused on functional connectivity without linking the activity in any processing area to specific cognitive or mental operations. To establish the latter connections, investigators rely on the techniques employed in more traditional neural imaging – identifying what types of cognitive tasks elicit changes in blood flow in particular brain regions. These findings, however, are now understood within the context of the networks in which the areas function. In one of the studies that first drew attention to the decrease in activity in the default network in task conditions, Andreasen et al. (1995) compared blood flow during an episodic-memory task with that produced both during a resting state and a semantic-memory task. They found that both the resting state and episodic-memory tasks generated higher BOLD levels than did the semantic-memory task in regions later identified as constituting the default network. In an attempt to figure out what was eliciting activity in these areas during the resting state, the researchers queried their subjects about what they were doing while lying in the scanner and found that they reported being engaged in 'a mixture of freely wandering past recollection, future plans, and other personal thoughts and experiences' – activities that plausibly draw upon episodic memory. This suggestion has been developed further by Buckner et al. (2008), who linked mental activity during the resting state to what Antrobus et al. (1970) characterized as *mind-wandering* and 'hypothesiz[ed] that the fundamental function of the default network is to facilitate flexible self-relevant mental explorations – simulations – that provide a means to anticipate and evaluate upcoming events before they happen' (Buckner et al. 2008, 2).

Although the approach to identifying and referring to cognitive operations being performed in brain regions begins in a manner comparable to more traditional approaches to neuroimaging, the strategies for further research go considerably beyond the traditional localizationist project. First, activities such as mind-wandering and self-relevant mental explorations are not being localized in a single brain region, but in a network of regions, and additional work is required to determine the specific operations performed by the different components of the network. The synchronization within the network then must be considered in characterizing each of these operations. Second, the coordination of the dynamic activities in different networks whose activity is anticorrelated also becomes an important issue. One of the interests in mind-wandering is that it often leads to subjects' failing to notice or attend to other stimuli. This may well be manifest in competitive interactions between different networks that modulate cognitive processing within each. The ability of network dynamics to alter processing has been demonstrated in a more local manner in an fMRI study which showed that the ongoing endogenous oscillation in motor areas could account for the variability in the BOLD signal recorded during motor behavior (button pressing) and moreover accounted for most of the variation in the strength of button presses (Fox et al. 2007).

In the previous section I discussed evidence for structural small-world organization in the brain, whereas in this section I have discussed comparable evidence for functionally organized networks. It is, of course, possible that these are unrelated, but that does not seem to be the case. Rather, the functionally characterized default network maps closely onto a structural network identified through diffusion tensor imaging tractography (Greicius et al. 2009). In addition to reviewing how, on a variety of measures, the small-world architectures found through structural and functional oscillations at low-frequencies tend to correspond, Bullmore and Sporns (2009) note how, at short temporal intervals there are dynamic functional changes, especially involving high-frequency oscillations, that are not reflected in changed structure. Nonetheless, the structural organization is likely to affect the faster dynamic processing and Bullmore and Sporns (2009, 196) identify several questions to address in further research:

> We need to understand more about the non-stationarity or metastability of brain functional networks. How does functional network topology change over time? Do functional networks exist in a dynamically critical state at some or all frequency intervals? What constraints on the itinerancy of network dynamics are imposed anatomically and how does the long-term history of functional activity in a network feed back on the development and remodelling of the anatomical connections between nodes?

Understanding processing in networks such as I have been describing requires adopting a different conception of a mechanism than one involving the sequential operation of nearly decomposable parts. Instead, it requires a perspective in which researchers characterize the components of mechanisms in terms of variables and represent the changes in values of these variables in terms of (differential) equations. One can then analyze the functioning of mechanisms in terms of the patterns of change over time in properties of their parts and operations, generating *dynamic mechanistic explanations* (Bechtel and Abrahamsen 2010). In developing equations to characterize the relations between operations performed by various parts, researchers are not limited to sequential operations, but are able to accommodate any dependency relation between parts, including those whereby activity elsewhere in a mechanism modulates the operation of a given part. The behavior of a small-world network can be extremely complex, and it is often only via simulation and the invocation of tools such as those of dynamical-systems theory (e.g. analyses of attractors in multidimensional phase spaces) that it is possible to anticipate how such mechanisms will behave. But unlike the dynamicists who oppose localization, researchers pursuing dynamic mechanistic accounts are still describing mechanisms. Even if the parts and especially the operations are not fixed but contextually modulated, they can still be characterized and referred to (e.g. in the equations developed to model their behavior).

The route to an adequate dynamic mechanistic explanation begins with strategies to localize operations in parts, where these are viewed as nearly decomposable constituents and assumed to operate in roughly a sequential manner. However, the results of such efforts only serve to characterize the parts and operations of the mechanism *to a first approximation* (Bechtel and Richardson [1993] 2010). Especially in biology, researchers increasingly expect that the parts and operations will be affected in a variety of ways by activity elsewhere in the mechanism, and as these additional interactions are identified, they need to be characterized in equations which can be used to understand the resulting complex dynamics.

IV CONCLUSIONS

The endeavor of localizing cognitive operations in the brain has traditionally been approached from the perspective of mechanistic explanation in which the regions of the brain are independent parts performing distinct information-processing operations. But research on both the structure and functional processes in the brain suggests a very different picture, in which the brain is organized as a small-world network. This still allows for local clusters to specialize in performing different operations and for researchers to refer to them, but the components are also regularly modulated by activity elsewhere in the brain. Such interactions do not entail a holism that defeats reference to localized components, but a perspective of a dynamically organized and modulated system in which the component operations are contextually modified by activity elsewhere. The initial assumption of a hierarchically organized nearly decomposable system must be modified, in the course of research, to take into account the sorts of modulations that occur in a small-world network. These modulating effects are represented in equations whose solutions reveal the temporal dynamics of the system. Within the context of such dynamic mechanistic explanations, localizing functions still plays an important role, but it is only an initial step and the resulting localization claims must be modified as researchers recognize how the whole mechanism functions in time, by modifying the operation of its own constituents. The references to parts and operations in the brain need to be couched within a dynamic perspective in which both the parts and operations change through time in complex ways.

REFERENCES

Achard, Sophie, Salvador, Raymond, Whitcher, Brandon, Suckling, John, and Bullmore, Ed (2006) A resilient, low-frequency, small-world human brain functional network with highly connected association cortical hubs. *Journal of Neuroscience* 26: 63–72.

Andreasen, Nancy C., O'Leary, Daniel S., Cizadlo, Ted, et al. (1995) Remembering the past: Two facets of episodic memory explored with positron emission tomography. *American Journal of Psychiatry* 152: 1576–1585.

Antrobus, John S., Singer, Jerome L., Goldstein, S., and Fortgang, M. (1970) Mindwandering and cognitive structure. *Transactions of the New York Academy of Sciences* 32: 242–252.

Bechtel, William (2008) *Mental Mechanisms*. London: Routledge.

Bechtel, William and Abrahamsen, Adele (2005) Explanation: A mechanist alternative. *Studies in History and Philosophy of Biological and Biomedical Sciences* 36: 421–441.

(2010) Dynamic mechanistic explanation: Computational modeling of circadian rhythms as an exemplar for cognitive science. *Studies in History and Philosophy of Science*, Part A, 41, no. 3: 321–333.

Bechtel, William and Richardson, Robert C. ([1993] 2010) *Discovering Complexity: Decomposition and Localization as Strategies in Scientific Research*. Cambridge, MA: MIT. Originally published by Princeton University Press in 1993.

Biswal, Bharat, Yetkin, F. Zerrin, Haughton, Victor M., and Hyde, James S. (1995) Functional connectivity in the motor cortex of resting human brain using echo-planar MRI. *Magnetic Resonance in Medicine* 34: 537–541.

Britten, Kenneth H., Shadlen, Michael N., Newsome, William T., and Movshon, J. Anthony (1992) The analysis of visual motion: A comparison of neuronal and psychophysical performance. *Journal of Neuroscience* 12: 4745–4765.

Broca, Paul (1861) Remarques sur le siége de la faculté du langage articulé, suivies d'une observation d'aphémie (perte de la parole). *Bulletin de la Société Anatomique* 6: 343–357.

Brodmann, Korbinian ([1909] 1994) *Vergleichende Lokalisationslehre der Grosshirnrinde*. Leipzig: J. A. Barth. Translated by L. J. Garvey as *Brodmann's Localization in the Cerebral Cortex: The Principles of Comparative Localisation in the Cerebral Cortex Based on Cytoarchitectonics*. New York: Springer, 2006.

Buckner, Randy L., Andrews-Hanna, Jessica R., and Schacter, Daniel L. (2008) The brain's default network. *Annals of the New York Academy of Sciences* 1124: 1–38.

Bullmore, Ed and Sporns, Olaf (2009) Complex brain networks: Graph theoretical analysis of structural and functional systems. *Nature Reviews Neuroscience* 10: 186–198.

Buzsáki, György and Draguhn, Andreas (2004) Neuronal oscillations in cortical networks. *Science* 304: 1926–1929.

Canolty, R. T., Edwards, E., Dalal, S. S., et al. (2006) High gamma power is phase-locked to theta oscillations in human neocortex. *Science* 313: 1626–1628.

Cordes, Dietmar, Haughton, Victor M., Arfanakis, Konstantinos, et al. (2000) Mapping functionally related regions of brain with functional connectivity MR imaging. *American Journal of Neuroradiology* 21: 1636–1644.

Felleman, Daniel J. and van Essen, David C. (1991) Distributed hierarchical processing in the primate cerebral cortex. *Cerebral Cortex* 1: 1–47.

Flourens, J. P. M. (1824) *Recherches expérimentales sur les propriétés et les fonctions du système nerveux dans les animaux vertébris*. Paris: Crevot.

Fox, Michael D., Snyder, Abraham Z., Vincent, Justin L., Corbetta, Maurizio, Van Essen, David C., and Raichle, Marcus E. (2005) The human brain is intrinsically organized into dynamic, anticorrelated functional networks. *Proceedings of the National Academy of Sciences of the United States of America* 102: 9673–9678.

Fox, Michael D., Snyder, Abraham Z., Vincent, Justin L., and Raichle, Marcus E. (2007) Intrinsic fluctuations within cortical systems account for intertrial variability in human behavior. *Neuron* 56: 171–184.

Fritsch, Gustav Theodor and Hitzig, Eduard (1870) Über die elecktrische Erregbarkeit des Grosshirns. *Arhiv für Anatomie und Physiologie* 37: 300–332.

Gong, Pulin and van Leeuwen, Cees (2004) Evolution to a small-world network with chaotic units. *Europhysics Letters* 67: 328–333.

Gray, Charles M. and Singer, Wolf (1989) Stimulus-specific neuronal oscillations in orientation columns of cat visual cortex. *Proceedings of the National Academy of Sciences of the United States of America* 86: 1698–1702.

Greicius, Michael D., Krasnow, Ben, Reiss, Allan L., and Menon, Vinod (2003) Functional connectivity in the resting brain: A network analysis of the default mode hypothesis. *Proceedings of the National Academy of Sciences of the United States of America* 100: 253–258.

Greicius, Michael D., Supekar, Kaustubh, Menon, Vinod, and Dougherty, Robert F. (2009) Resting-state functional connectivity reflects structural connectivity in the default mode network. *Cerebral Cortex* 19: 72–78.

Hagmann, Patric, Cammoun, Leila, Gigandet, Xavier, et al. (2008) Mapping the structural core of human cerebral cortex. *PLoS Biology* 6: e159.

Hubel, David H. and Wiesel, Torsten N. (1962) Receptive fields, binocular interaction and functional architecture in the cat's visual cortex. *Journal of Physiology* 160: 106–154.

(1968) Receptive fields and functional architecture of monkey striate cortex. *Journal of Physiology* 195: 215–243.

Ingvar, David H. (1975) Patterns of brain activity revealed by measurements of regional cerebral blood flow. In David H. Ingvar and Niels A. Lassen (eds.) *Brain Work: The Coupling of Function, Metabolism, and Blood Flow in the Brain: Proceedings of the Alfred Benzon Symposium VIII, Copenhagen, 26–30 May 1974, Held at the Premises of the Royal Danish Academy of Sciences and Letters, Copenhagen.* New York: Academic, pp. 397–413.

Kandel, Eric R. (1976) *Cellular Basis of Behavior: An Introduction to Behavioral Neurobiology.* San Francisco: W. H. Freeman.

Lakatos, Peter, Shah, Ankoor S., Knuth, Kevin H., Ulbert, Istvan, Karmos, George, and Schroeder, Charles E. (2005) An oscillatory hierarchy controlling neuronal excitability and stimulus processing in the auditory cortex. *Journal of Neurophysiology* 94: 1904–1911.

Lashley, Karl S. (1950) In search of the engram. In Society for Experimental Biology (ed.) *Physiological Mechanisms in Animal Behaviour*, Symposia of the Society for Experimental Biology IV. New York: Academic Press; Cambridge University Press, pp. 454–482.

Laufs, Helmut, Krakow, K., Sterzer, P., et al. (2003) Electroencephalographic signatures of attentional and cognitive default modes in spontaneous brain activity fluctuations at rest. *Proceedings of the National Academy of Sciences of the United States of America* 100: 11053–11058.
Llinás, Rodolfo R. (1988) The intrinsic electrophysiological properties of mammalian neurons: Insights into central nervous system function. *Science* 242: 1654–1664.
Llinás, Rodolfo R. and Yarom, Y. (1981) Properties and distribution of ionic conductances generating electroresponsiveness of mammalian inferior olivary neurones *in vitro*. *Journal of Physiology* 315: 569–584.
Lorente de Nó, Rafael (1938) Analysis of the activity of the chains of internuncial neurons. *Journal of Neurophysiology* 1: 207–244.
Machamer, Peter, Darden, Lindley, and Craver, Carl F. (2000) Thinking about mechanisms. *Philosophy of Science* 67: 1–25.
Mantini, D., Perrucci, M. G., Del Gratta, C., Romani, G. L., and Corbetta, Maurizio (2007) Electrophysiological signatures of resting state networks in the human brain. *Proceedings of the National Academy of Sciences of the United States of America* 104: 13170–13175.
Margulis, Lynn [see also Sagan, Lynn] (1981) *Symbiosis in Cell Evolution: Life and Its Environment on the Early Earth*. San Francisco: W. H. Freeman.
Milgram, Stanley (1967) The small world problem. *Psychology Today* 2: 60–67.
Milner, A. David and Goodale, Melvyn G. (1995) *The Visual Brain in Action*. Oxford University Press.
Munk, Hermann (1881) *Über die Funktionen der Grosshirnrinde*. Berlin: A. Hirschwald.
Posner, M. I. and Raichle, M. E. (1994) *Images of Mind*. San Francisco: W. H. Freeman.
Raichle, Marcus E., MacLeod, Ann Mary, Snyder, Abraham Z., Powers, William J., Gusnard, Debra A., and Shulman, Gordon L. (2001) A default mode of brain function. *Proceedings of the National Academy of Sciences of the United States of America* 98: 676–682.
Raichle, Marcus E. and Mintun, Mark A. (2006) Brain work and brain imaging. *Annual Review of Neuroscience* 29: 449–476.
Rubinov, Mikail, Sporns, Olaf, van Leeuwen, Cees, and Breakspear, Michael (2009) Symbiotic relationship between brain structure and dynamics. *BMC Neuroscience* 10: 55.
Sagan, Lynn [Lynn Margulis] (1967) On the origin of mitosing cells. *Journal of Theoretical Biology* 14: 255–274.
Scannell, J. W., Burns, G. A. P. C., Hilgetag, C. C., O'Neil, M. A., and Young, Malcolm P. (1999) The connectional organization of the cortico-thalamic system of the cat. *Cerebral Cortex* 9: 277–299.
Simon, Herbert A. (1962) The architecture of complexity: hierarchic systems. *Proceedings of the American Philosophical Society* 106: 467–482.
Song, Sen, Sjöström, Per Jesper, Reigl, Markus, Nelson, Sacha, and Chklovskii, Dmitri B. (2005) Highly nonrandom features of synaptic connectivity in local cortical circuits. *PLoS Biology* 3: e68.

Sporns, Olaf and Zwi, Jonathan D. (2004) The small world of the cerebral cortex. *Neuroinformatics* 2: 145–162.

Steriade, Mircea and Llinás, Rodolfo R. (1988) The functional states of the thalamus and the associated neuronal interplay. *Physiological Reviews* 68: 649–742.

Ungerleider, Leslie G. and Mishkin, Mortimer (1982) Two cortical visual systems. In D. J. Ingle, M. A. Goodale, and R. J. W. Mansfield (eds.) *Analysis of Visual Behavior*. Cambridge, MA: MIT Press, pp. 549–586.

Uttal, William R. (2001) *The New Phrenology: The Limits of Localizing Cognitive Processes in the Brain*. Cambridge, MA: MIT Press.

van Essen, David C. and Gallant, Jack L. (1994) Neural mechanisms of form and motion processing in the primate visual system. *Neuron* 13: 1–10.

van Orden, Guy C. and Paap, Kenneth R. (1997) Functional neural images fail to discover the pieces of the mind in the parts of the brain. *Philosophy of Science* 64: S85–S94.

van Orden, Guy C., Pennington, Bruce F., and Stone, Gregory O. (2001) What do double dissociations prove? Inductive methods and isolable systems. *Cognitive Science* 25: 111–172.

Watts, Duncan and Strogatz, Steven (1998) Collective dynamics of small worlds. *Nature* 393: 440–442.

Wimsatt, William C. (1986) Forms of aggregativity. In A. Donagan, N. Perovich, and M. Wedin (eds.) *Human Nature and Natural Knowledge*. Dordrecht: Reidel, pp. 259–293.

Young, Malcolm P. (1993) The organization of neural systems in the primate cerebral cortex. *Proceedings of the Royal Society of London, Series B: Biological Sciences* 252: 13–18.

Index

285

www.ingramcontent.com/pod-product-compliance
Ingram Content Group UK Ltd.
Pitfield, Milton Keynes, MK11 3LW, UK
UKHW020436180125
453697UK00006B/69